Electronic Signal Conditioning

Electronic Signal Conditioning

Bruce Newby

Butterworth-Heinemann Ltd
Linacre House, Jordan Hill, Oxford OX2 8DP

A member of the Reed Elsevier plc group

OXFORD LONDON BOSTON
MUNICH NEW DELHI SINGAPORE SYDNEY
TOKYO TORONTO WELLINGTON

First published 1994

British Library Cataloguing in Publication Data
Newby, B. W. G.
  Electronic Signal Conditioning
  I. Title
  621.382

ISBN 0 7506 1844 2

Library of Congress Cataloguing in Publication Data
Newby, B. W. G. (Bruce W. G.)
  Electronic signal conditioning/Bruce W. G. Newby.
  p.   cm.
  Includes bibliographical references and index.
  ISBN 0 7506 1844 2
  1. Electronic circuits.   I. Title.
  TK7867.N447   93–51084
  621.3815–dc20   CIP

Typeset by TecSet Ltd, Wallington, Surrey
Printed and bound in Great Britain by Clays Ltd, St Ives plc

Contents

Preface

The dictionary describes a *signal* as 'a transmitted effect conveying a message'; *conditioning* is defined as 'the putting into the required state'. In the field of electronic engineering, the combined term of *signal conditioning* is taken as meaning the manipulation of a voltage or current waveform into a more useful or manageable size or shape or its undergoing a logical or mathematical operation.

The aim of this book is to collect together and explain some of the more common conditioning processes to which electronic signals may be subjected. It has been written especially for BTEC second year ONC, HNC and HND students as well as for those in the first year of a relevant engineering degree. The book is particularly suitable for study programmes which include any of the following modules: analogue electronics, mechatronics, instrumentation, control, signal conditioning and the like.

Because operational amplifiers are much used in signal conditioning systems, Chapter 2 is devoted to a convenient review of these devices and concludes with an explanation of the rather novel method of measuring their performance in *bits of accuracy*. In general, the depth of treatment in Chapter 2 is sufficient for most of the relevant courses but recommended further reading is the book *Operational Amplifiers*, third edition, by Clayton and Newby and published by Butterworth-Heinemann. This gives a more advanced mathematical appreciation of operational amplifiers and their applications.

The first ten chapters of the present book are mainly concerned with analogue conditioning techniques and the last two introduce digital conditioning. Chapter 11 deals with the conversion of signals between analogue and digital form. Chapter 12 is an introduction to further aspects of digital signal conditioning. Programmable logic controllers are discussed in some detail. Digital filters are introduced using a simple qualitative approach thus avoiding the welter of mathematics usually associated with these devices.

Finally, Texas Instruments are thanked for their kind permission to include unabridged copies of their technical data sheets in Appendices 1 and 2.

Bruce Newby

1

D.c. and a.c. bridges

1.1 Introduction

The requirement to condition a signal typically arises from a system like that outlined in Figure 1.1. The sensor block receives a physical stimulus in the form of heat, light, sound, movement or the like and it converts this stimulation into an electrical signal which is related to the magnitude of the stimulation. The output from the sensor block (which is usually a transducer) may, for example, be required to drive a meter or other display system. The output from the sensing transducer is unlikely to be sufficiently powerful to operate the display system without first being amplified. The function of amplification is in this case the necessary 'signal conditioning' to change the transmitted effect (voltage waveform) conveying the message (about the stimulus) by putting it into the required state (sufficiently powerful) to drive the subsequent display or record function block.

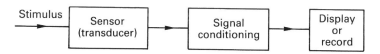

Figure 1.1 *Block diagram of a simple instrumentation system*

More specific examples of the 'signal conditioning' function are illustrated in Figure 1.2. In (a) the amplifier simply conditions the weak signal from the microphone by increasing its power sufficiently to drive the tape recorder. In (b) the signal conditioning function is shown as two distinct steps. The rough d.c. pulses produced by the tachogenerator at the rate of, say, one pulse per revolution of the shaft, are firstly 'squared-up' in the shaping circuit and then differentiated to produce the necessary sharp 'pips' for the counter to register. The number of pips counted in a second gives the angular velocity of the shaft in revolutions per second (rev/s).

The above are only two examples of what signal conditioning entails. Subsequent chapters will examine other electronic signal conditioning functions which frequently are required in the operation of practical engineering systems. However, before proceeding further, it is important that the reader is made familiar with the use of d.c. and a.c. bridges for the coupling of the sensed signal into the subsequent signal conditioning block.

Suppose that we have a transducer which will sense a physical stimulus by assuming an electrical resistance related to the magnitude of the applied stimulus be it heat, light, movement or the like. The simple potential divider (or

2 D.c. and a.c. bridges

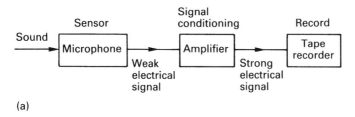

(a)

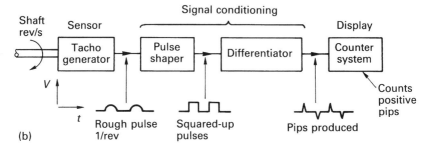

(b)

Figure 1.2 *Examples of the need for signal conditioning in (a) an audio system and (b) measuring the angular velocity of a rotating shaft*

half bridge) circuit shown in Figure 1.3 would produce an output voltage, e_o, given by the following expression:

$$e_o = e_i R_1/(R_1 + R_2)$$

R_1 is the sensor and R_2 a fixed resistance; e_i is the input voltage. A problem with this simple circuit can arise because there is a voltage output even when the stimulus is zero. Some instrumentation and control circuit applications require a zero output for zero input situation. Interchanging the positions of R_1 and R_2 would not cure the problem of a continuous output because the sensor transducer has a finite value of resistance (its passive resistance) even

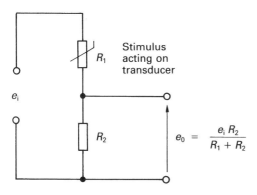

Figure 1.3 *Half bridge circuit for sensing a physical stimulus*

when not stimulated. A solution is to use a pair of similar half bridges in parallel and to take the output from between the two bridge midpoints. The circuit arrangement is, in effect, the well-known d.c. Wheatstone bridge usually used for resistance measurement. In this case, however, the 'unknown resistance' is the sensor transducer which when stimulated causes the bridge circuit to produce an output voltage related to the amount of applied stimulation.

1.2 Wheatstone's d.c. bridge

Figure 1.4 shows a typical Wheatstone bridge configuration using a strain gauge of passive resistance 120 Ω as the sensing element in one arm of the bridge and fixed resistors also of 120 Ω resistance in each of the other three arms. For the output, e_o, from the bridge to be zero the voltage at X must be equal and opposite to the voltage at Y.

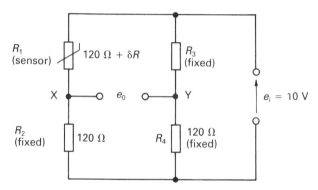

Figure 1.4 *Wheatstone's bridge using a strain gauge sensor*

The voltage at X, V_X, is given by

$$V_X = e_i R_2 / (R_1 + R_2) \tag{1.1}$$

and similarly, the voltage at Y, V_Y, by

$$V_Y = e_i R_4 / (R_3 + R_4) \tag{1.2}$$

Therefore, e_o must be the difference between V_X and V_Y and if e_o is to be zero, which it will be if the bridge circuit is in balance, then V_X must equal V_Y.
Clearly then,

$$R_2 / (R_1 + R_2) = R_4 / (R_3 + R_4) \tag{1.3}$$

This is known as the 'balance equation' for the bridge circuit and gives the relationship between the resistances in the four arms of the bridge for zero output voltage.
Now suppose that the strain gauge, R_1, is subjected to a mechanical movement which places it in tension. This will cause the gauge resistance to increase by a small amount δR. The effective value of R_1 will become $(R_1 + \delta R)$

and so will reduce the value of V_X such that it no longer equals unchanged V_Y. The difference between the two will be the unbalanced bridge output voltage.

A typical value for δR for a 120 Ω strain gauge is 0.05 Ω. This means that the strained value of R_1 becomes 120.05 Ω and assuming the values given in Figure 1.4 for the other resistors and the d.c. supply, substituting these in Equations 1.1 and 1.2 gives

$$V_X = 10 \times 120/(120.05 + 120) = 4.99895855 \text{ V}$$

and

$$V_Y = 10 \times 120/(120 + 120) = 5.00000000 \text{ V}$$

The difference is $e_o = 1.04145$ mV.

Note that with e_i being a direct voltage in the sense shown, point Y is positive relative to point X, the out of balance current flowing from Y to X.

The above bridge arrangement for producing a measurement signal uses what is called the 'null method' for adjusting the bridge output to zero, that is, for initially balancing the bridge. Further, the output produced by imbalance can be measured by a centre-zero meter and so not only can the size of the stimulation be measured, but also can its sense or direction. A more practical arrangement for facilitating these functions is shown in Figure 1.5. The variable resistor R_4 is adjusted for zero meter output with the sensor not stimulated. Initially, R_5 is set at its highest resistance to prevent any large out of balance currents damaging the meter. As the balance conditions are approached, the value of R_5 can be reduced to zero, effectively increasing the sensitivity of the meter.

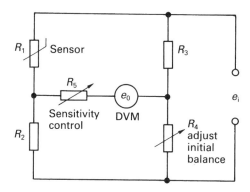

Figure 1.5 *More practical bridge circuit to facilitate initial balance and meter sensitivity control. DVM, digital voltmeter*

Figure 1.6 shows a bridge arrangement using two sensitive elements. If these are positioned as shown, and acted upon by the same stimulus, the output from the bridge is enhanced.

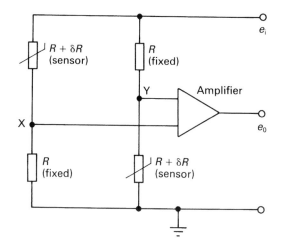

Figure 1.6 *An increased output can be obtained by using two sensitive bridge elements*

Exercise

The circuit shown in Figure 1.6 is connected to a supply battery of 9 V. The two sensors are resistance strain gauges of passive resistance 120 Ω. Two high precision fixed resistors are used; both of them are 120 Ω in value. If the amplifier has a gain of 8 and if both of the strain gauges are stimulated to suffer a 0.1% resistance change, show that the output from the circuit, e_o, is 36 mV.

It is worth noting at this point that while the above examples show voltage outputs in the order of several millivolts, there will be cases when a small stimulus will produce an output of less than one millivolt. In this situation, the noise voltage spikes generated by the measurement circuit itself may be sufficiently large partially to hide the wanted signal voltage. The problem for the subsequent signal conditioning then can become one of noise suppression as well as that of amplification. The subject of noise is discussed further in Chapter 8.

While the Wheatstone bridge circuit is satisfactory for use with d.c. circuits which use only resistive elements, it is not so useful when reactive components are required to be part of the bridge circuit. This is because an inductor appears as a short circuit and a capacitor as an open circuit to a.c. For these frequency dependent components there are hundreds of a.c. bridge circuits available.

1.3 A.c. bridge circuits

The requirement for balance in any bridge circuit is that there is *never* a potential difference between X and Y in Figure 1.4. This means that the

6 D.c. and a.c. bridges

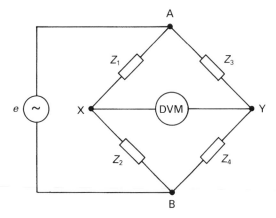

Figure 1.7 *General case for an a.c. bridge containing four reactive elements. DVM, digital voltmeter*

voltages at X and Y must not only be of the same magnitude, but they must also be in phase, that is peak at the same time. With the Wheatstone bridge using pure resistors and a d.c. energising supply, we are concerned only with magnitude; there are no phase differences around the circuit. Even with an a.c. supply, because the Wheatstone bridge uses pure resistors, the values of which do not change with frequency, the bridge can still be balanced by the adjustment of the voltage magnitudes at X and Y provided an a.c. sensitive null detector is used.

If we now wish to use a bridge circuit with a sensor which contains a frequency sensitive element (a reactance) then we run into problems when trying initially to balance the bridge before applying the stimulus to the sensor. We now must balance both the *magnitude* and the *phase* of the voltages at X and Y for a zero response on the null detector. Figure 1.7 shows the general case of an a.c. bridge employing complex impedances, Z, in each of the four arms and where:

$Z = R + jX_{\mathrm{L}}$ (for a partly inductive impedance) .

and

$Z = R - jX_{\mathrm{C}}$ (for a partly capacitive impedance)

X_{L} is the inductive reactance and is equal to $2\pi fL$ Ω and X_{C} is the capacitive reactance and is equal to $1/2\pi fC$ Ω, where L is the value of the inductance in henry, C is the capacitance in farad, f is the a.c. frequency in hertz, and R is the resistance in ohms. The mathematical operator $+j$ means add vectorially after rotating 90° anticlockwise, while $-j$ means the same thing but involving an initial clockwise rotation. From this it can be shown that $|Z| = \sqrt{(R^2 + X^2)}$ Ω.

In order to establish the balance equation for Figure 1.7 we can apply Equation 1.3, but using impedances instead of pure resistances, to equate the voltages at X and Y:

$$Z_2/(Z_1 + Z_2) = Z_4/(Z_3 + Z_4)$$

Therefore,

$$Z_2(Z_3 + Z_4) = Z_4(Z_1 + Z_2)$$

and

$$Z_2Z_3 + Z_2Z_4 = Z_1Z_4 + Z_2Z_4$$

Finally,

$$\mathbf{Z_1 Z_4 = Z_2 Z_3} \tag{1.4}$$

Equation 1.4 gives the full condition of balance.

Suppose all the impedances are of the form $R_1 + jX_1$, $R_2 + jX_2$, etc. This means

$$(R_1 + jX_1)(R_4 + jX_4) = (R_2 + jX_2)(R_3 + jX_3)$$

or

$$R_1R_4 + jX_4R_1 + jX_1R_4 - X_1X_4 = R_2R_3 + jX_2R_3 + jX_3R_2 - X_2X_3$$

Collecting in-phase (real) and out-of-phase (imaginary) terms,

$$R_1R_2 - X_1X_4 + j(X_4R_1 + X_1R_4) = R_2R_3 - X_2X_3 + j(X_2R_3 + X_3R_2)$$

and for balance the real terms on each side of the equation must equate as must the imaginary terms. Therefore,

$$R_1R_2 - X_1X_4 = R_2R_3 - X_2X_3 \text{ (real terms)}$$

and

$$X_4R_1 + X_1X_4 = X_2R_3 + X_3R_2 \text{ (imaginary terms)}$$

Clearly, attempting to adjust all of the resistances and reactances which could appear in each of the four arms of the bridge would make the task of achieving a complete balance extremely tedious.

In practice the problem is simplified by making two of the impedances, say Z_1 and Z_2, pure resistances so that both X_1 and $X_2 = 0$. The equations for balancing the real terms and then the imaginary terms become:

$$R_1R_4 = R_2R_3 \tag{1.5}$$

and

$$R_1X_4 = R_2X_3 \tag{1.6}$$

Now suppose that Z_4 is an unknown impedance and Z_3 is a variable impedance of which we can adjust both R_3 and X_3 in order to achieve bridge balance. The usual way of obtaining balance is to apply the a.c. supply and then adjust R_3 for a minimum reading on the null detector. Then adjust X_3 to further reduce the out of balance reading. Alternate adjustments of R_3 and X_3 are continued, possibly increasing the sensitivity of the detector as the balance settings are approached, until, ideally, a null reading is noted. The final values of R_3 and

X_3, together with the previously known values of R_1 and R_2, are substituted into Equations 1.5 and 1.6 and the values of R_4 and X_4 are calculated.

The bridge circuit may not necessarily be used for the measurement of an unknown impedance but for the production of an electrical signal related to the change of impedance caused to Z_4 which in practice may be an inductive or capacitive transducer.

Depending upon the nature of the unknown impedance to be measured, there are hundreds of a.c. bridge circuit designs from which to choose the most suitable. Each design has its merits. Some give the best results for measuring a large inductance, some for a small capacitance, some for a high frequency a.c. supply and others for high Q circuits, and so on. Many of these bridges are available commercially and their manufacturers usually give precise operating instructions as to their use for the best results. Some of the bridges have been cleverly designed such that, despite using frequency sensitive reactances, the balance condition is independent of the a.c. frequency used.

1.3.1 The Maxwell bridge

The Maxwell bridge is a useful arrangement for determining the value of an inductance of appreciable resistance all of which is regarded as being in series. The circuit is shown in Figure 1.8. The unknown inductance, or the inductive sensor in an instrumentation role, is represented by Z_4. If the condition for initial balance of the circuit is analysed, it will be seen that the component parts of Z_4, R_4 and L_4 are largely determined by the adjustable values of C_1 and R_1 and are independent of the a.c. frequency used.

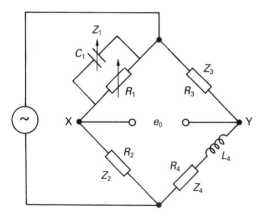

Figure 1.8 *The Maxwell bridge*

The balance equation is obtained by equating the voltages at X and Y, V_X and V_Y respectively, to produce a zero output voltage, e_o:

$$V_X = V_Y$$

This produces the balance equation (Equation 1.4):

$$Z_1 Z_4 = Z_2 Z_3$$

In this case,

$$Z_1/(1 + j\omega R_1 C_1)$$

$$Z_2 = R_2,$$

$$Z_3 = R_3,$$

$$Z_4 = R_4 + j\omega L_4$$

Substituting these complex impedances into the balance equation gives:

$$[R_1/(1 + j\omega R_1 C_1)](R_4 + j\omega L_4) = R_2 R_3$$

So

$$R_1(R_4 + j\omega L_4) = R_2 R_3(1 + j\omega R_1 C_1)$$

Equating the real part of the left-hand side of the equation with that of the right-hand side gives:

$$R_1 R_4 = R_2 R_3$$

whence

$$\boldsymbol{R_4 = R_2 R_3 / R_1}$$

Equating the imaginary parts gives:

$$\omega L_4 R_1 = \omega R_1 C_1 R_2 R_3$$

The ωR_1 on each side of the equation cancels (showing frequency independence) and the result is:

$$\boldsymbol{L_4 = R_2 R_3 C_1}$$

It will be noted that the inductance, L_4, can be determined without the provision of an expensive standard adjustable inductance; adjustable resistors and capacitors are more readily available.

1.3.2 The Hay bridge

The circuit for this bridge is shown in Figure 1.9. This is similar to the Maxwell bridge but is more suitable for measuring inductances having only a small series resistance, that is for high Q coils. Any resistance, R_4, is regarded as being in parallel with the true inductance, L_4. A similar analysis of the balance conditions shows that this circuit is also frequency independent and $R_1 R_4 = R_2 R_3$ while $L_4 = R_2 R_3 C_1$.

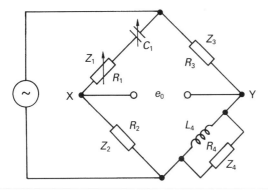

Figure 1.9 *The Hay bridge*

1.3.3 The Schering bridge

The circuit for this bridge is shown in Figure 1.10. The bridge can be used for measuring very small capacitances such as the interelectrode capacitances of electronic devices as well as the value of ordinary capacitors. Also, it can be used for measuring the loss resistance at high voltage. C_1 should be a high quality variable capacitor and R_1 an adjustable resistor. C_3 and R_2 need not be special. The unknown capacitor, C_4, is assumed to have its losses represented by a series resistance, R_4.

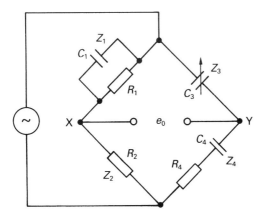

Figure 1.10 *The Shering bridge*

Analysis of the balance conditions shows that $R_1C_3 = R_2C_4$ and $R_4C_4 = R_1C_1$.

1.3.4 The Wien frequency bridge

This special circuit is arranged such that the bridge balances to provide a zero output at only one frequency of the applied a.c. This means that the Wien bridge can be used as a notch filter to block a particular frequency for some possible signal conditioning requirement. The circuit is shown in Figure 1.11. The notable points are that not only are the impedances Z_3 and Z_4 pure resistances, but Z_4 is twice the value of Z_3. Further, both the variable resistances are the same value and are mechanically ganged such that they both adjust simultaneously to a value R. The fixed capacitors C_1 and C_2 are also the same value, C.

$$Z_1 = R/(1 + j\omega CR)$$

$$Z_2 = (R + 1/j\omega C)$$

$$Z_3 = R_3$$

$$Z_4 = 2R_3$$

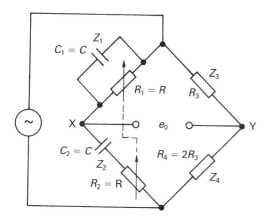

Figure 1.11 *The Wien frequency bridge*

Substituting these values in the balance equation $Z_1 Z_4 = Z_2 Z_3$ and equating real and imaginary parts in the normal way results in the necessary balance frequency, f Hz, being given by the expression:

$$\omega = 1/CR \quad \text{rad/s}, \quad \text{where} \quad \omega = 2\pi f$$

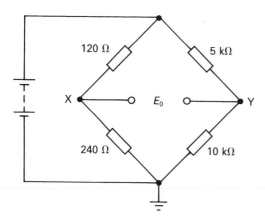

Figure 1.12 *Circuit for Exercise 1.1*

Exercises

1.1 (a) Is the circuit shown in Figure 1.12 in balance? What is the value of e_o?

(b) If in the above circuit the 10 kΩ resistor were replaced by a 12 kΩ resistor, calculate the value of E_o and state which of the points X and Y is the more positive.

1.2 For the Maxwell bridge shown in Figure 1.8, write down the balance equation and develop this to show that when e_o is zero,

$$R_4 = R_2 R_3 / R_1 \quad \text{and} \quad L_4 = R_2 R_3 C_1.$$

1.3 In Figure 1.8, suppose that the circuit is in balance when $C_1 = 65$ pF, $R_1 = R_2 = 12$ kΩ, $R_3 = 120$ kΩ and the frequency of the a.c. supply is 10 kHz. Calculate the values of R_4, L_4 and Z_4.

1.4 For the circuit shown in Figure 1.9, show that when e_o is zero it is not necessary to know the frequency of the a.c. supply to determine R_4 and L_4.

1.5 The signal conditioning section of an instrumentation system requires a tunable filter at its input in order to reject selected spot frequencies in the range 1 to 3 kHz. The circuit chosen to undertake this filtering task is the Wien frequency bridge, as shown in Figure 1.11, having component values as follows: $C_1 = C_2 = 0.01\,\mu$F and $R_3 = 100$ kΩ. State a suitable value for R_4 and calculate the higher and lower values of R_1 and R_2 necessary to achieve the required frequency coverage.

1.6 A transducer sensor having a passive impedance comprising a capacitance of 0.1 μF in series with a resistance of 500 Ω is interfaced to a signal conditioning system by a Shering bridge as shown in Figure 1.10. With no stimulus applied to the sensor the bridge is in balance and the output voltage, e_o, is zero. The bridge component values at balance are as follows: $C_1 = 0.025$ μF, $C_3 = 0.1$ μF, $R_1 = R_2 = 2$ kΩ, R_4 and C_4 together comprise the sensor, and the a.c. supply is 10 V at a frequency of 1 kHz. If the sensor is stimulated, so causing its capacitance to increase by 10%, calculate the bridge output voltage.

2

A review of operational amplifiers

2.1 Introduction

In subsequent chapters of this book the operational amplifier plays a prominent part in the many signal conditioning circuits considered. For this reason alone, it is important that the reader is given a fairly comprehensive review of the techniques involved when using these very common devices in a signal conditioning role.

The term *operational amplifier or op-amp* was originally used by workers in the analogue computing industry to describe electronic devices which would undertake a variety of mathematical operations. In simple terms, the devices were very high gain d.c. voltage amplifiers each fitted with a different feedback circuit between its output and input to produce a specific mathematical function. The functions obtainable included simple amplification, differentiation, integration, addition, subtraction and the like. Many of the functions undertaken by the early analogue computers have now become the province of the ubiquitous digital computer, causing a large decline in the numbers of the former. But the operational amplifier has not suffered the same fate; it has been retained and further developed and now is extensively used in the analogue electronics field of instrumentation.

While operational amplifiers can be built from discrete components, the present proliferation of these devices has been largely brought about by their ready availability commercially in modular and integrated form. Many manufacturers now market a very wide range of high performance operational amplifiers which by the addition of a few simple external components can be configured to undertake specific signal processing tasks. Not only can they be designed to undertake the usual mathematical operations, but they can also be arranged to simulate, for example, the performance of a proposed mechanical, hydraulic or electrical control system. This can be a valuable cost saving exercise since system problems are encountered and designed out before the expensive hardware assembly stage is started.

It is not necessary for the average user to understand completely the internal workings of the operational amplifier. But it is important that the functioning of externally fitted feedback circuits is well understood, and in order to help in this aim we shall first examine the attributes of the ideal operational amplifier.

The circuit symbols used with operational amplifiers are shown in Figure 2.1. The basic operational amplifier has two inputs and outputs. If one of the pair of input or output terminals is either earthed or connected (commoned) to the other then the terminals are said to be *single ended*. If pairs of terminals are left unconnected they are said to be *floating*. When the pair of input terminals are floating it is the voltage difference between the two which constitutes the true

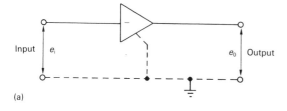

(a)

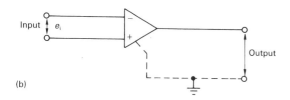

(b)

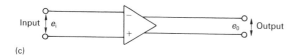

(c)

Figure 2.1 *Circuit symbols for the three operational amplifier configurations: (a) single ended input and output (the dotted earth return is often omitted); (b) differential input, single ended output; (c) differential input and output*

input signal. Of the three circuit configurations shown, the differential input and single ended output shown in Figure 2.1(b) is the more common.

The two input terminals are labelled 'positive' and 'negative'. This has nothing to do with electrical polarity but rather indicates the relative phases of the input signal and output signal it produces. A signal applied to the 'negative' input terminal appears inverted at the output; that is, 180° out of phase. Similarly, a signal applied to the 'positive' input terminal appears at the output in phase with the input signal. The two input terminals, labelled '−' and '+', are therefore called the *inverting* and the *non-inverting* terminals respectively.

Most operational amplifiers have the differential input arrangement because this allows more feedback flexibility than does the single ended input. In fact, the output of the perfect or ideal differential input operational amplifier depends only on the difference between the voltages on the two input terminals and the nature of the particular feedback circuit being used. Any common voltage that the two input terminals may have with respect to earth is called a *common mode* input voltage. Since a common mode input voltage applied to both inputs produces no differential input signal, there is no common mode output signal. This is a very important attribute because it means that unwanted noise voltages applied simultaneously to both of the input terminals, being 'common mode' voltages, have no effect on the output. This all assumes that the operational amplifier is ideal and perhaps it would be appropriate at this point to clarify what is meant by an ideal operational amplifier.

2.2 The ideal operational amplifier

2.2.1 The characteristics of an ideal operational amplifier

The ideal operational amplifier is taken as having five characteristics as follows.

- *Infinite gain*   This makes the performance of the amplifier dependent only on the nature of the external components which have been added to form a feedback circuit.
- *Infinite input impedance*   This ensures that no current flows into the amplifier input terminals.
- *Zero output impedance*   This makes the amplifier capable of supplying an infinitely large output current without its performance being affected.
- *Infinite bandwidth*   This ensures that the amplifier will respond equally to all input signals varying in frequency from zero (d.c.) to infinitely high. This also means that the ideal operational amplifier will have a zero response time before an input signal appears at the output. Therefore there will be no phase difference between the amplifier input and output signals.
- *Zero voltage and current offset*   This means that no matter what the source impedance connected to the operational amplifier input terminals may be, when the input voltage to the amplifier is zero then the output voltage will likewise be zero.

2.2.2 Operational feedback

In order to make our basic very high gain operational amplifier perform the different operations we may require of it, we apply feedback. Using the differential input, single ended output configuration, there are two basic methods of achieving this feedback. Figure 2.2 shows the circuits for these two methods. In both cases the signal fed back through R_f is *voltage derived* (from the output voltage), but is applied in *shunt* with e_i, in the case of the inverting amplifier configuration, and effectively in *series* with e_i, in the non-inverting case. In both cases the voltage feedback is applied to the amplifier inverting input terminal and, because of the phase inversion through the amplifier, the feedback is negative.

The negative voltage feedback thus applied to the inverting terminal of the amplifier differential input ensures that any difference input signal, e, tends to be cancelled out. In other words, sufficient of the output voltage caused by e in the first place is fed back negatively to reduce e to zero. This is a very important result and is often stated as one of the two *golden rules* governing the functioning of ideal operational amplifiers:

- *Golden rule 1*   When negative feedback is applied to an ideal operational amplifier the inverting and non-inverting input terminals are forced to the same voltage.

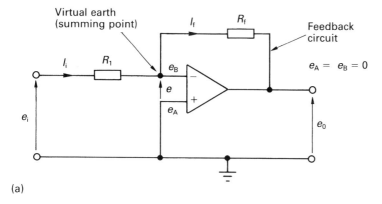

(a)

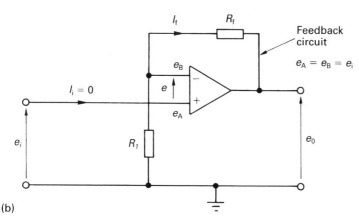

(b)

Figure 2.2 *The two basic operational amplifier circuits: (a) inverting; (b) non-inverting*

The second golden rule for understanding ideal operational amplifier action stems from its having an infinite input impedance. In Figure 2.2(a) this means that no current can flow into the input terminals of the amplifier. Any input current, I_i, passing through the input resistor, R_1, must of necessity all flow through the feedback resistor, R_f.

- *Golden rule 2* No current flows into either input terminal of an ideal operational amplifier.

The application of these two rules to Figure 2.2(a) produces two further aspects:

- If the non-inverting terminal of the amplifier is connected to earth, then Golden Rule 1 means that the inverting terminal is also at a *virtual earth*.
- If the inverting input terminal is at a virtual zero potential it can be used as a common junction for several parallel current inputs with their sources adversely interacting; it is a *summing point*.

We can now use these rules to derive expressions for the voltage gain of the circuits shown in Figure 2.2 as follows.

2.2.3 Inverting operational amplifier voltage gain

The circuit is shown in Figure 2.2(a). With e_A being earthed, Golden Rule 1 dictates that e_B must also be of zero potential. Therefore, the input current, I_i, flowing through R_1 is given by:

$$I_i = e_i/R_1$$

Golden Rule 2 effectively says that $I_i = I_f$, in which case since

$$I_f = (e_B - e_o)/R_f = -e_o/R_f$$

we can write

$$e_i/R_1 = -e_o/R_f$$

Rearranging this, the circuit gain e_o/e_i is given by $-R_f/R_i$.

The gain of the amplifier block without any feedback is called the amplifier *open loop gain*, A_{VOL}. In the case of the ideal operational amplifier this is infinitely high. The gain of the circuit complete with its feedback loop, as derived above, is traditionally called its *closed loop gain*, A_{VCL}.

So the inverting amplifier circuit gain is completely independent of the infinite open loop gain of the ideal amplifier block itself and is determined only by the components comprising the feedback network.

$$A_{VCL} = -R_f/R_i$$

2.2.4 Non-inverting operational amplifier voltage gain

The relevant circuit is shown in Figure 2.2(b). Once again, there is no input current flow into the non-inverting input terminal and its voltage is $e_A = e_i$. With $e = 0$, we note that

$$e_B = e_i \tag{2.1}$$

The output voltage, e_o, is generated by the current I_f which flows as indicated from earth through R_1 and R_f to the amplifier output terminal. Therefore,

$$I_f = (0 - e_o)/(R_1 + R_f)$$
$$= -e_o/(R_1 + R_f) \tag{2.2}$$

However, we can also use Equation 2.1 to deduce that

$$I_f = (e_i - e_o)/R_f \tag{2.3}$$

By equating Equations 2.2 and 2.3 and rearranging them to make e_o/e_i the subject, we obtain the equation for the closed loop gain of the non-inverting amplifier as follows:

$$A_{VCL} = 1 + R_f/R_i$$

The two equations for the closed loop gains are different in sign to correspond with the inverting and non-inverting roles, but other than that, for ratios of R_f/R_i greater than 10, the amplification factor for the two circuits is much the same. The major difference between the two circuit configurations is the effective input resistance which they present to a signal source.

2.2.5 Inverting and non-inverting amplifier input impedances

The signal input terminal of the ideal inverting operational amplifier, with shunt voltage negative feedback and its other ínput earthed, is itself at virtual earth. Therefore, the amplifier block presents a virtual zero impedance to the signal, e_i. In Figure 2.2(a), the input current, I_i, to the circuit is limited only by R_1 which represents the whole circuit input impedance as seen by the signal source. So by applying shunt negative feedback the infinitely high input impedance of the amplifier block itself has, by the addition of the feedback network, been effectively reduced to zero.

The opposite effect on the whole circuit input impedance arises with the ideal non-inverting operational amplifier configuration. The input signal, e_i, applied to the amplifier block non-inverting input terminal is effectively cancelled by the series voltage feedback to its inverting terminal. The overall result is that no current flows into the circuit and the signal source is presented with an infinitely high impedance.

2.2.6 The operational amplifier as a current-to-voltage converter

Figure 2.3 illustrates how this is achieved using the standard inverting amplifier connection. Point X is held at earth potential by the action of shunt negative feedback. Any input current is therefore forced to flow through the feedback resistor to the amplifier output terminal. The voltage drop caused by I_i flowing through R_f produces the output voltage e_o where

$$e_o = -I_i R_f$$

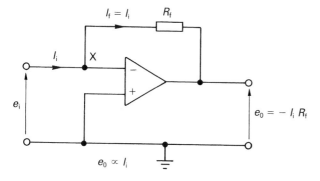

Figure 2.3 *The operational amplifier current-to-voltage converter*

So the output voltage is simply the input current times the scaling factor of R_f. This circuit can be used as an ideal ammeter where the current to be measured is passed through $R_f(I_i = I_f)$, and the amplifier block by taking no current itself does not load and therefore does not affect the circuit under test.

2.2.7 The ideal operational amplifier for summing currents or voltages

Figure 2.4(a) shows the circuit arrangement for the summing of independent currents. The currents I_1, I_2 and I_3 are all connected to the virtual earth at point X. No current can flow into the amplifier inverting terminal so the sum of the

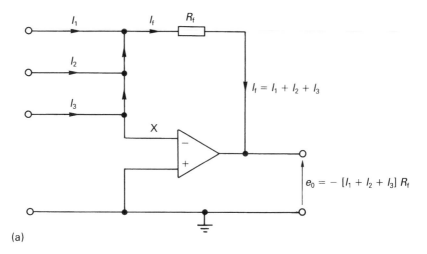

(a)

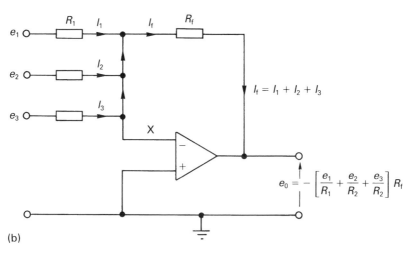

(b)

Figure 2.4 *Operational amplifier circuits for the summation of: (a) currents; (b) voltages*

input currents flows through R_f. The volt drop so caused across R_f is in fact the output voltage, e_o:

$$e_o = -(I_1 + I_2 + I_3)R_f$$

Figure 2.4(b) shows a similar arrangement, but with the addition of resistors in the input lines, which enables the output voltage to be proportional to the sum of the three independent input voltages:

$$e_o = -R_f[(e_1/R_1) + (e_2/R_2) + (e_3/R_3)]$$

2.2.8 The ideal operational amplifier as a voltage-to-current converter

The circuit for this function is shown in Figure 2.5. The input voltage, e_i, is fed to the non-inverting terminal and the usual operational amplifier action to make e_i also appear at point X follows. The feedback current required to maintain the amplifier differential input voltage at zero is I_L:

$$I_L = e_i/R_1$$

It will be noted that the value of I_L is dependent only on the value of R_1 and has nothing to do with the load resistor, R_f.

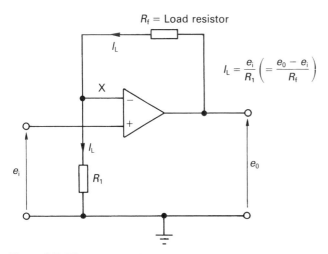

Figure 2.5 *The operational amplifier voltage-to-current converter. The load current is proportional to the input voltage*

2.2.9 The ideal operational amplifier as a buffer stage

With the full output voltage fed back to the inverting terminal as shown by the circuit in Figure 2.6, the input voltage, e_i, appears at the output without any voltage amplification having occurred. But the ideal amplifier block has an infinite input impedance and a zero output impedance and this is just the

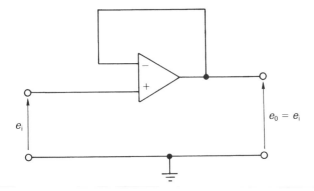

Figure 2.6 *The operational amplifier unity gain buffer amplifier. High input impedance, low output impedance*

requirement for heavy current amplification or for acting as an isolating buffer between a high impedance and a low impedance.

2.2.10 The ideal operational amplifier as a differential subtractor

The circuit is shown in Figure 2.7. Note how for the best results the input and feedback resistor values are chosen.

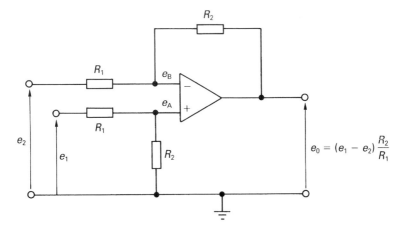

Figure 2.7 *The operational amplifier subtractor*

The value of e_A at the amplifier non-inverting input is simply given by the potential divider across e_1.

$$e_A = e_1 R_2 / (R_1 + R_2)$$
$$= e_1 R_2 / R_T \quad \text{(where } R_T = R_1 + R_2\text{)}$$

E_B = (Volts due to e_2 with e_o earthed) + (Volts due to e_o with e_2 earthed)

$$= e_2 R_2 / R_T + e_o R_1 / R_T$$

But the normal operational amplifier action makes $e_A = e_B$ so that we can write:

$$e_1 R_2 / R_T = e_2 R_2 / R_T + e_o R_1 / R_T$$

Cancelling R_T on both sides and rearranging gives:

$$\boldsymbol{e_o = (e_1 - e_2)R_2 / R_1}$$

This shows that the output voltage is proportional to the difference between the two input voltages.

2.2.11 The ideal operational amplifier as an integrator

Should we take the standard inverting operational amplifier circuit and replace the feedback resistor with a capacitor, then the response of the circuit is changed. A positive voltage step input appears at the output as a negatively sloping ramp output. This is brought about by the normal capacitor action of resisting any immediate change to the voltage across it. In other words, we have produced a circuit the output of which is the integral of its input. The appropriate circuit is shown in Figure 2.8. As ever with the ideal amplifier, the circuit input current is the same as that flowing through the feedback component which in this case is a capacitor.

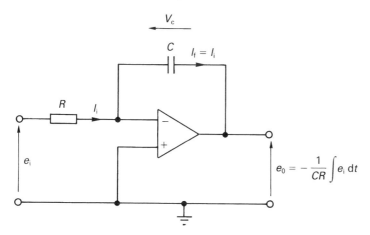

Figure 2.8 *The operational amplifier*

The instantaneous charging current flowing into a capacitor is given by the general equation

$$i = C \, \mathrm{d}V/\mathrm{d}t$$

where V is the voltage across the initially discharged capacitor which has been charging for time t.

Applying this equation to our particular situation we can write that, since $I_i = I_f$:

$$\frac{e_i}{R} = c\frac{\mathrm{d}V_c}{\mathrm{d}t}$$

But the voltage across the capacitor is equal in magnitude but opposite in sign to the output voltage, e_o. Therefore,

$$\frac{e_i}{R} = -C\frac{\mathrm{d}e_o}{\mathrm{d}t}$$

and

$$e_o = -\frac{1}{CR}\int e_i \, \mathrm{d}t$$

The output is seen to be proportional to the time integral of the input voltage.

2.2.12 The ideal operational amplifier as a differentiator

Figure 2.9 shows how by interchanging the positions of the capacitor and resistor we can change the integrator into a differentiator.

Since the normal ideal operational amplifier action makes the voltage at

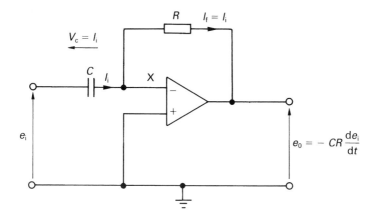

Figure 2.9 *The operational amplifier differentiator*

point X zero and $I_i = I_f$, we can write:

$$I_i = C \, \mathrm{d}e_i/\mathrm{d}t = -e_o/R$$

which after rearrangement gives

$$e_o = -CR \, de_i/dt$$

The output voltage is proportional to the differential coefficient, with respect to time, of the input voltage.

2.3 The practical operational amplifier

Today, there is little need to design and construct an operational amplifier using discrete components. There are thousands of ready made devices in integrated circuit form which can be configured to suit different applications by the addition of external feedback components. Circuit layout design is made simpler because the many different manufacturers tend to use standard packaging and pin numbering. All manufacturers produce data sheets which explain the characteristics of their operational amplifiers. In many cases the data sheets also include comprehensive notes covering circuit design and useful practical applications for the device. (Specimen manufacturers' data sheets are given in Appendix 1.)

The practical amplifier does not perform in the ideal way that we have assumed thus far in this chapter. Some of the differences between the ideal and the practical situations are indicated in Table 2.1. Because it does not have an infinite input impedance, the practical operational amplifier will in reality take a small amount of current into its inverting and non-inverting terminals. Also, being basically a differential amplifier with two inputs, it is not possible to arrange for the circuitry to remain in precise balance under all conditions and hence there tends to be an output present in the absence of a true input signal. The limited open loop gain and bandwidth together with the existence of a small amount of output resistance and the requirement for bias currents all generally degrade the performance of the practical operational amplifier. The remainder of this chapter will be devoted to a brief discussion of these shortcomings. A good starting point is the analysis of a circuit which is more representative of the practical operational amplifier than those used thus far.

Table 2.1 *Comparison of ideal and practical operational amplifier characteristics*

Characteristic	Ideal	Typical
Voltage gain	Infinite	200 000
Input resistance	Infinite	2 MΩ
Output resistance	Zero	75 Ω
Bandwidth	Infinite	10 kHz
Input offset volts	Zero	1 mV

2.3.1 Analysis of the practical operational amplifier follower

The circuit configuration is shown in Figure 2.10 and now the finite input impedance, Z_1, is taken into account. The input signal voltage, e_i, is applied directly to the non-inverting amplifier terminal which causes the input current, I_1, to flow into Z_1. An in-phase output voltage, e_o, is also produced and a fraction, β, of this is fed back through R_f to the amplifier inverting terminal. The voltage appearing at the inverting terminal is e_f and this equals βe_o. The effective input voltage producing e_o is in fact the voltage differential, e, across the two input terminals of the amplifier, given by:

$$e = e_i - e_f$$

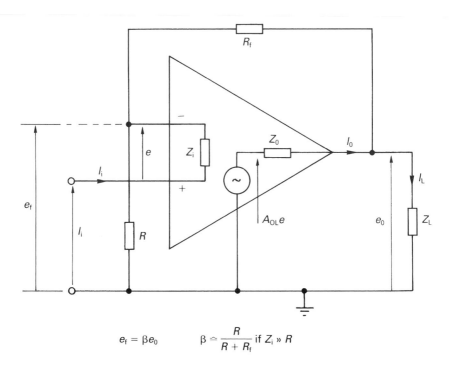

$$e_f = \beta e_0 \qquad \beta \simeq \frac{R}{R + R_f} \text{ if } Z_i \gg R$$

Figure 2.10 *Practical non-inverting (follower) amplifier – series voltage feedback case*

It is this voltage, e, multiplied by the amplifier open loop gain, A_{OL}, which produces an effective signal voltage of eA_{OL}. Unfortunately, this voltage does not appear at the amplifier output terminals without first being attenuated. This is caused by the volt drop produced by the output current, I_o, flowing through the amplifier output impedance, Z_o.

The output of the operational amplifier appears to act as a voltage generator of e.m.f. eA_{OL} and internal impedance Z_o, producing an output terminal voltage of e_o. The equation for this is:

$$e_o = eA_{OL} - I_oZ_o$$

Replacing e by $(e_i - e_f) = (e_i - \beta e_o)$ we can write:

$$e_o = A_{OL}(e_i - \beta e_o) - I_oZ_o$$
$$= e_iA_{OL} - e_o\beta A_{OL} - I_oZ_o$$

Rearranging,

$$e_o(1 + \beta A_{OL}) = e_iA_{OL} - I_oZ_o$$

whence:

$$e_o = \frac{e_iA_{OL}}{1 + \beta A_{OL}} - \frac{I_oZ_o}{1 + \beta A_{OL}}$$

Note how the first term of this equation represents the e.m.f. generated by the operational amplifier (it is less than eA_{OL} because of the series voltage negative feedback) while the second term is the internal volt drop across an output impedance reduced from the original Z_o. These two terms form the closed loop circuit performance parameters as follows:

$$A_{CL} = \frac{A_{OL}}{1 + \beta A_{OL}} \quad \text{(the closed loop gain)}$$

$$Z_{oCL} = \frac{Z_o}{1 + \beta A_{OL}} \quad \text{(the output impedance)}$$

Note that if the term βA_{OL} is made very large by using a very high gain amplifier, in the rearranged equation for the closed loop gain, namely

$$A_{CL} = \frac{1}{\beta}\left(\frac{1}{1 + \dfrac{1}{\beta A_{OL}}}\right)$$

the term in parentheses is almost equal to unity. Thus the value of the closed loop gain, A_{CL}, is shown to be virtually decided by the value of $1/\beta$ alone.

The term βA_{OL} is seen to have considerable influence on the values of both the amplifier closed loop gain and its output impedance. The term βA_{OL} is given a special name: the *loop gain*.

Having seen how negative feedback affects the gain and output impedance of the practical follower amplifier, let us now see what happens to the input impedance which without negative feedback is simply Z_1 in Figure 2.10.

We already know that $e_i = e + e_f = e + \beta e_o$. We can see from Figure 2.10 that provided $R_f \gg R_L$ (and we shall ensure that it is) then the output voltage, e_o, is developed across Z_L in series with Z_o and the e.m.f. eA_{OL}. Therefore,

$$e_o = eA_{OL}\frac{Z_L}{Z_o + Z_L}$$

Substitution for e_o gives

$$e_o = e + e\beta A_{OL}\frac{Z_L}{Z_o + Z_L}$$

and rearranging,

$$e_o = e\left(1 + \beta A_{OL}\frac{Z_L}{Z_o + Z_L}\right)$$

Now the closed loop input impedance is $Z_{iCL} = e_i/I_i$. Therefore,

$$Z_{iCL} = \frac{e}{I_i}\left(1 + \beta A_{OL}\frac{Z_L}{Z_o + Z_L}\right)$$

But since $e/I_i = Z_i$ we can write

$$Z_{iCL} = Z_i\left(1 + \beta A_{OL}\frac{Z_L}{Z_o + Z_L}\right)$$

In other words, the open loop impedance is increased by negative series voltage feedback and the major influence in the increase is again the open loop gain, βA_{OL}.

2.3.2 Analysis of the practical operational amplifier inverter

Figure 2.11 shows the relevant circuit. The situation is a little more complicated than for the follower because our calculations now have to take into account

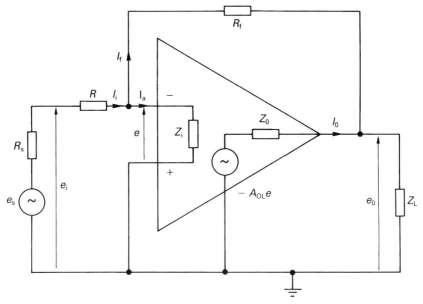

Figure 2.11 *Practical inverting amplifier – shunt voltage feedback case*

the internal resistance of the signal source. Once again we shall take the output voltage, e_o, as being produced by the amplification of the differential input voltage, e. The generated output e.m.f. is $-A_{OL}e$; the negative sign arising because the signal is fed into the inverting input terminal. The final output voltage, e_o, is given by:

$$e_o = -(A_o e - I_o Z_o)$$

Now e, the differential input voltage, is simply the combined effect of e_s and e_o at the amplifier inverting input terminal. We calculate e by superimposing the individual effects of e_s and e_o with the other shorted to earth:

$$e = \text{The effect of } e_s + \text{The effect of } e_o$$

$$e = e_s \frac{R_f}{(R + R_s + R_f)} + e_o \frac{R + R_s}{(R + R_s + R_f)}$$

Substituting this value of e in the previous equation for e_o we have:

$$e = -A_{OL} e_s \frac{R_f}{(R + R_s + R_f)} - \beta A_{OL} e_o + I_o Z_o$$

$$\text{where } \beta = \frac{R + R_s}{(R + R_s + R_f)}, \text{ and}$$

$$e_o = -A_{OL} e_s (1 - \beta) - \beta A_{OL} e_o + I_o Z_o$$

Rearranging,

$$e_o(1 + \beta A_{OL}) = -A_{OL} e_s (1 - \beta) + I_o Z_o$$

Dividing both sides of this equation by $e_s(1 + \beta A_{OL})$ we obtain the equivalent 'generator equation' for e_o from which we can obtain the following:

$$A_{CL} = -\frac{A_{OL}(1 - \beta)}{(1 + \beta A_{OL})}$$

and

$$Z_{oCL} = \frac{Z_o}{(1 + \beta A_{OL})}$$

We can rearrange the equation for A_{CL} by replacing $(1-\beta)$ by $R_f/(R + R_s + R_f)$ as follows:

$$A_{CL} = -\frac{R_f}{(R + R_s + R_f)}\frac{A_{OL}}{(1 + \beta A_{OL})}$$

$$= \frac{R_f}{R + R_s + R_f)}\frac{1}{\beta}\left(\frac{A_{OL}}{\dfrac{1}{\beta} + A_{OL}}\right)$$

$$= -\frac{R_f}{R + R_s}\left(\frac{1}{1 + \dfrac{1}{\beta A_{OL}}}\right)$$

Provided that the loop gain, βA_{OL}, is very high and that $R_s \ll R$, then the closed loop gain of the operational amplifier is given by the value of the feedback components, that is it equals $1/\beta$. This is a similar result to that obtained for the follower amplifier case in the previous section.

If we now have a look at the effect the shunt voltage feedback has on the input impedance we find that this is decreased rather than increased as it was with series voltage feedback. Referring to Figure 2.11 we see that the input current, I_i, now divides at the inverting input terminal into I_f flowing round the feedback loop and I_a flowing into the amplifier open loop internal input impedance, Z_i. Therefore,

$$I_i = I_f + I_a$$

$$= \frac{e - e_o}{R_f} + \frac{e}{Z_i}$$

$$= \frac{e}{R_f} + \frac{e A_{OL}}{R_f} + \frac{e}{Z_i}$$

$$= e\left(\frac{1}{Z_i} + \frac{1}{R_f} + \frac{A_{OL}}{R_f}\right)$$

$$= e\left(\frac{1}{Z_i} + \frac{1 + A_{OL}}{R_f}\right)$$

Now the reciprocal of the closed loop input impedance is

$$\frac{1}{Z_{iCL}} = \frac{I_i}{e} = \frac{1}{Z_I} + \frac{1 + A_{OL}}{R_f}$$

Therefore the open loop input impedance has effectively been reduced by its being shunted by an additional impedance equal to $R_f/(1 + A_{OL})$. This is shown in Figure 2.12.

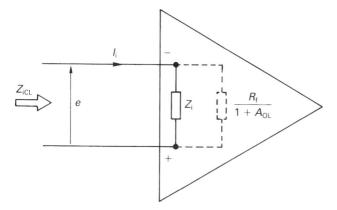

Figure 2.12 *The effect of negative shunt voltage feedback on the inverting operational amplifier. From the viewpoint of the source, the input impedance has been reduced from Z_i to $Z_i \parallel \frac{R_f}{1 + A_{OL}}$*

2.3.3 Negative current feedback

The two previous sections have dealt with the effects on the practical operational amplifier of voltage negative feedback. The purpose of voltage negative feedback is to stabilise the output voltage and so it is derived directly from the output voltage. The purpose of negative current feedback is to stabilise the output current, and hence the feedback signal (albeit a voltage) is derived from the output current. The way this is done is to pass the output current through a resistor. The voltage across this resistor is taken as the feedback voltage and applied back to the operational amplifier inverting input terminal. Figure 2.13

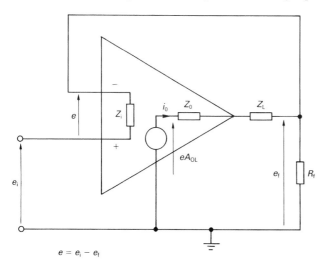

$$e = e_i - e_f$$

Figure 2.13 *Series current feedback*

shows a circuit arrangement which achieves series current feedback because the input signal, e_i, and the feedback voltage, e_f, are effectively applied to the amplifier differential input terminals in series opposition.

The highlights of the analysis of the circuit in Figure 2.13 follow:

$$i_o = \frac{eA_{OL}}{Z_o + Z_L + R_f}$$

But $e = e_i - e_f$ and, if we assume that all of i_o flows through r_f because of $R_f \ll Z_i$, then $e_f = i_o R_f$.

So we can substitute $e = (e_i - i_o R_f)$ in the above equation, and after re-arrangement obtain:

$$i_o = \frac{A_{OL} e_i}{Z_o + Z_L + (1 + A_{OL})R_f}$$

This means that effectively the circuit output impedance has been increased by the addition of the term $R_f A_{OL}$.

The equation for i_o can be rearranged to give the now familiar form:

$$i_o = \frac{e_i}{R_f} \left(\frac{1}{1 + \dfrac{1}{\beta A_{OL}}} \right)$$

where in this case the voltage feedback fraction is $\beta = R_f/(R_f + Z_L + Z_o)$, which has a value dependent upon the load impedance.

The importance of the term βA_{OL} is again apparent. The expression in parentheses, namely $[1/(1 + 1/\beta A_{OL})]$, is frequently called the *gain error factor*. If βA_{OL} is large, the gain error factor approaches unity and the output current is virtually determined by the value of the input voltage and that of the feedback resistor.

2.4 Determining the values for β and e_o from circuit component values

When determining the values for β from a circuit the general formula to use is:

$$\beta = Z_p/(Z_p + Z_f)$$

where Z_p is the parallel sum of all the impedance paths from the amplifier inverting input terminal, through the signal sources, to earth, and Z_f is the impedance of the feedback loop from the amplifier output terminal to the amplifier inverting input feedback fraction. The feedback fraction so calculated can be used to predict the gain of a practical amplifier, bearing in mind that

Actual gain = Ideal gain $\times$ Gain error factor

or

$$A_{CL} = 1/\beta \frac{1}{1 + 1/\beta A_{OL}}$$

For example, suppose we have the circuit arrangement as shown in Figure 2.14 then the feedback fraction is given by:

$$\beta = \frac{1}{1 + \dfrac{R_f(R_1 + R_2)}{R_1 R_2}}$$

The expression for the calculation of the output voltage,

$$e_0 = -\left[e_1 \left(\frac{R_f}{R_1} \right) + e_2 \left(\frac{R_f}{R_2} \right) + e_3 \left(\frac{1}{\beta} \right) - e_4 \left(\frac{1}{\beta} \right) \right] \left(\frac{1}{1 + \dfrac{1}{\beta A_{OL}}} \right)$$

basically comprises the sum of the different input signals after their having undergone *ideal* amplification and then been all multiplied by the *gain error factor* to give the actual output voltage.

Note that e_1 and e_2 undergo an *ideal signal gain*, R_f/R_1 and R_f/R_2 respectively, while e_3 and e_4 both suffer the *ideal closed loop gain*, $1/\beta$.

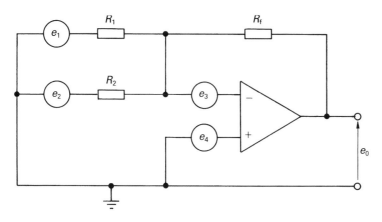

$$\beta = \frac{R_1//R_2}{(R_1//R_2) + R_f} = \frac{R_1 R_2/(R_1 + R_2)}{[R_1 R_2/(R_1 + R_2)] + R_f} = \frac{1}{1 + [R_f(R_1 + R_2)/R_1 R_2]}$$

$$e_0 = -\left[e_1 \left(\frac{R_f}{R_1} \right) + e_2 \left(\frac{R_f}{R_2} \right) + e_3 \left(\frac{1}{\beta} \right) - e_4 \left(\frac{1}{\beta} \right) \right] \left(\frac{1}{1 + \dfrac{1}{\beta A_{OL}}} \right)$$

Figure 2.14 *Determination of expressions for feedback fraction and output voltage*

2.5 Operational amplifier frequency response

Our discussion thus far has not mentioned the time the output of an operational amplifier takes to respond to an input signal. While it is convenient to assume that an input of e volts applied to the input of an inverting amplifier will produce an immediate $-eA_{OL}$ volts at its output, there is in practice a short time delay before the output reaches this value. For d.c. and low frequency a.c. input signals, the time delay in the response of the operational amplifier may not be significant. However, for high frequency signal inputs or transient (d.c. voltage step) inputs, the time delay between input and output response will cause phase differences which may have to be taken into account. Also, large amplitude input signals can cause output distortion because of the amplifier being driven into saturated conduction. Therefore, in order to simplify matters we shall consider the amplifier response to only small amplitude sinusoidal inputs.

2.5.1 Gain versus frequency response and Bode plots

The usual way of illustrating the variation of an amplifier's output versus changes in the frequency of a constant magnitude input signal is to plot *gain* against the *logarithm to the base ten of the frequency*. Rather than expressing the gain as the ratio of the output and input voltages, e_o/e_i, it is more common to express the gain in decibels (dB). The relationship between the gain in decibels and the gain as a voltage ratio, e_o/e_i, is:

Voltage gain (dB) $= 20 \log_{10}(e_o/e_i)$

A few examples of the use of this equation are shown in Table 2.2. Note how a negative decibel rating indicates an attenuation, whereas a negative ratio would still mean amplification but with a phase inversion. It is important to appreciate that power is proportional to (voltage)^2 and hence the significance of the remarks in the final column of the table.

Table 2.2   *Voltage ratio and decibel relationship*

e_o/e_i	dB	*Power output*
0.1	-20	
0.5	-6	Quarter
0.707	-3	Half
1	0	Unity
1.41	3	Double
2	6	
10	20	
100	40	

The variation of amplifier output with different input frequencies involves not only changes in output magnitude but also changes in its phase relationship with the input. It is therefore important that the reader has an appreciation of the 'operator j' notation method of representing both the magnitude and the phase of complex quantities. Figure 2.15 is a brief reminder of how the complex number system works.

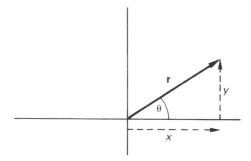

The vector of magnitude |r| and phase angle θ
can be represented using the 'operator j' notation by

$$\mathbf{r} = x + \mathrm{j}\, y$$

where $|\mathbf{r}| = \sqrt{x^2 + y^2}$, $\Theta = \tan^{-1}y/x$, $\mathrm{j} = \sqrt{-1}$

Figure 2.15 *Complex number representation*

The gain versus frequency diagram, or response curve, for an amplifier is usually obtained by supplying the input from a constant voltage but variable frequency source. The input frequency is varied in steps from zero frequency (d.c.) up to the higher frequency point where the output voltage has fallen to equal the input voltage, that is to the *unity gain* point. However, not only is there a variation of gain with frequency, but there is also a phase lag between the output and input. This enables us to plot two diagrams as shown in Figure 2.16, one for gain and the other for phase lag against log frequency. The frequency is plotted on a logarithmic scale for convenience; it helps to keep the frequency scale to manageable proportions. The dotted curves in Figure 2.16 represent the practical response while the solid straight lines are known as the *Bode plot approximations*. Let us now take a more detailed look at the Bode plot system.

Most operational amplifiers are designed to produce a gain variation with frequency in accordance with the equation:

$$A_{\mathrm{OL}(\mathrm{j}f)} = \frac{A_{\mathrm{OL}}}{1 + \mathrm{j}\dfrac{f}{f_{\mathrm{c}}}}$$

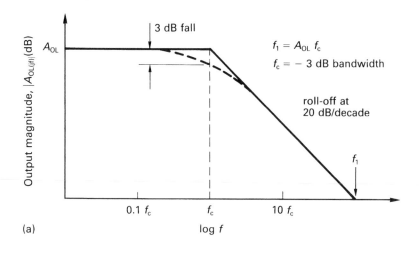

(a)

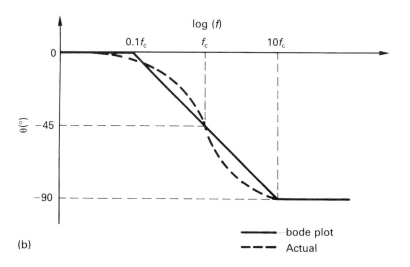

(b)

Figure 2.16 *Variation of operational amplifier output with frequency: (a) output magnitude; (b) output phase lag*

where $A_{OL(jf)}$ is a complex quantity which represents both the magnitude and phase lag of the gain at any particular frequency f, A_{OL} represents the amplifier d.c. gain, and f_c is a designed constant called the *break frequency*.

The magnitude of the response, as obtained from the above equation, is:

$$| A_{OL(jf)} | = \frac{A_{OL}}{\sqrt{\left[1^2 + \left(\dfrac{f}{f_c} \right)^2 \right]}}$$

The phase angle is given by:

$$\theta = -tan^{-1} f/f_c$$

Figure 2.16(a) is the diagram obtained from the equation for the gain magnitude. For values of $f \ll f_c$ the gain is A_{OL}. However, as f increases from zero, there is a progressive decrease in the open loop gain which, at about the frequency marked as f_c, turns into a marked decline. At frequencies higher than f_c the curve tends to a straight line negative slope of 20 dB/frequency decade. This linear *roll-off* continues down to cross the gain = 0 dB axis at a frequency marked as f_1. This is called the *unity gain frequency*.

The Bode diagram for the phase change through the amplifier is depicted in Figure 2.16(b). The phase change is very low for low frequency signals, a 45° lag appears at f_c and at frequencies higher than about $10f_c$ the phase change settles down to a steady 90°. The dotted line shows the smooth curve obtained by plotting θ, as calculated from the tangent equation, against log f. The solid line represents the Bode plot approximation.

The method of constructing the Bode plot straight lines is quite simple if A_{OL} and f_1 are both known. Usually, they can be obtained straight from the operational amplifier manufacturer's data sheet. A horizontal line is drawn at the value given for A_{OL} and a second straight line of negative slope of 20 dB/ decade through the unity gain frequency, f_1. The intersection of the two straight lines defines the break frequency f_c. It will be noted that the break frequency is where the gain has fallen by 3 dB or, in other words, where the amplifier output power is only half its d.c. rating.

Because two or more Bode diagrams, drawn with the gain in decibels and the frequency to a log scale, can be added together graphically quite easily, they are very useful in predicting the overall gain and possible instability of cascaded amplifier stages.

2.6 More operational amplifier terms

In the previous sections we have seen the way in which the gain and phase responses of practical operational amplifiers vary from the ideal. Provided we know what the variations are we can make allowances accordingly. Some additional practical terms are now listed, more in the spirit of alerting the reader to their existence, rather than providing any in-depth discussion about them.

The idealised transfer characteristic for an operational amplifier is shown in Figure 2.17. The output voltage, e_o, is shown as being a linear function of the differential input, $(e_1 - e_2)$, up to the point where saturation of the amplifier occurs. This is where the output voltage cannot exceed the positive and

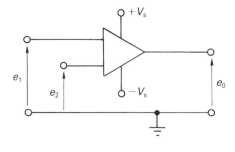

(a)

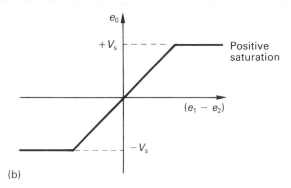

(b)

Figure 2.17 *The operational amplifier in open loop mode: (a) the circuit; (b) the idealised transfer characteristic*

negative values of the amplifier d.c. power supply. This helps to explain the first two terms listed below.

- *Maximum output voltage swing*   The peak output voltage which can occur without the amplifier running into saturated conduction so causing clipping of the output voltage waveform.
- *Maximum voltage between inputs*   The maximum permitted differential input voltage without damage being caused to the amplifier.
- *Maximum common mode voltage*   The maximum permitted common mode voltage which may be applied to both inputs without damaging the amplifier.
- *Phase margin*   The amount by which the excess phase shift (phase shift over and above the built-in 180° required for negative feedback and obtained by returning the feedback signal to the inverting input terminal) is less than 180° at that frequency at which the magnitude of the loop gain is unity. It is required to prevent the amplifier producing unwanted oscillations.
- *CMRR (common mode rejection ratio)*   This output voltage ratio is a measure of how well the amplifier will reject common mode (noise) voltage inputs compared with its willingness to amplify a differential voltage signal applied to its inputs.

- *Amplifier slew rate* A measure of how quickly the output of the amplifier will follow a voltage step input. It is usually measured in volts per microsecond (V/μs).
- *Bias current and input voltage offset* Together these are necessary to ensure that the output of the operational amplifier is zero when there is no signal input.

These and other terms used with operational amplifiers are shown in the specimen manufacturers' data sheets given in Appendix 1.

2.7 Operational amplifier accuracy expressed in 'bits'

The real operational amplifier is not ideal in many ways. The necessary input bias currents flowing through the input resistors together with the unwanted input offset voltage cause a d.c. output error. Voltage and current noise sources limit resolution and the dynamic range. The roll-off of the open loop gain limits the accuracy at the higher signal frequencies and the finite slew rate affects the output settling times after the application of an input change. Changes in the operating temperature and the power supplied also have an effect.

The different error effects which limit the ideal accuracy of the operational amplifier may be combined mathematically into a single number of merit. This number can be related to the accuracy of digital systems in the form of *bits of accuracy*. (The representation of a number by binary digits is dealt with in more detail in Section 11.2.4.)

The extent of the improvement in the performance of an operational amplifier depends upon the difference in magnitude between the open loop and closed loop gain factors. As shown in Figure 2.18, at low frequencies the difference $(1 + \beta A_o)$ is large and the output voltage is close to ideal; at frequencies above the break frequency, $(1 + \beta A_o)$ decays with the open loop gain roll-off and becomes 0 dB (unity gain) where the open loop and closed loop gain curves intersect. Thus, at the higher frequencies the operational amplifier input and output impedances are degraded and therefore it is always wise to limit the upper operational frequency to a point where $(1 + \beta A_o)$ is still relatively large.

The accuracy of an operational amplifier is typically limited by two important parameters: the non-zero offsets at low frequencies, and the reduced open loop gain at high frequencies. The gain error is calculated by taking the difference in magnitude between the ideal output voltage and the practical output. The ratio of the error voltage to the maximum possible output voltage, V_{max}, is then converted into *bits* using the equation:

$$\text{Bit accuracy} = -\log_2\left(\frac{\text{Error}}{V_{max}}\right) - 1$$

Now the main factor limiting the amplifier accuracy at low frequencies is the input offset voltage, V_{IO}, and if this happened to be 2 mV and the amplifier had a closed loop gain of 10, the error voltage would be $10 \times 2\text{ mV} = 20\text{ mV}$. If

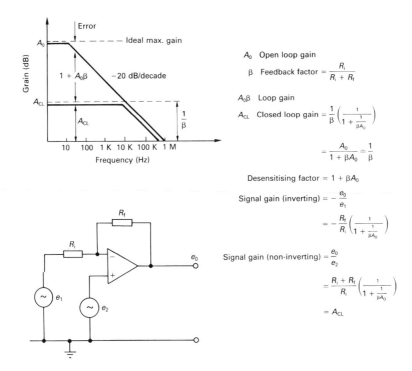

Figure 2.18 *Operational amplifier – summary*

V_{max} were 18 V, the bit accuracy using the above equation would be 8.8 bits; for $V_{IO} = 0.5$ mV the accuracy would be 10.8 bits, and so on. As the frequency is increased from d.c. the output starts to fall and where it has fallen to $V_{max}/2$, the error will be $0.5V_{max}$ and the accuracy will be virtually 0 bits.

Exercises

2.1   For each of the circuits shown in Figure 2.19, calculate the output voltage e_o.

2.2   For each of the two circuits shown in Figure 2.19, calculate the output voltages 3 ms after the application of the input voltages.

2.3   A signal conditioning system requires an inverting amplifier stage with a signal gain of 50 and an input resistance of 10 kΩ. It is decided to use an operational amplifier for the task.

   (a) Assuming an ideal operational amplifier, suggest suitable values for the input and feedback resistors.

   (b) If a practical operational amplifier, having a finite open loop gain of only 25 000, is used, calculate the percentage

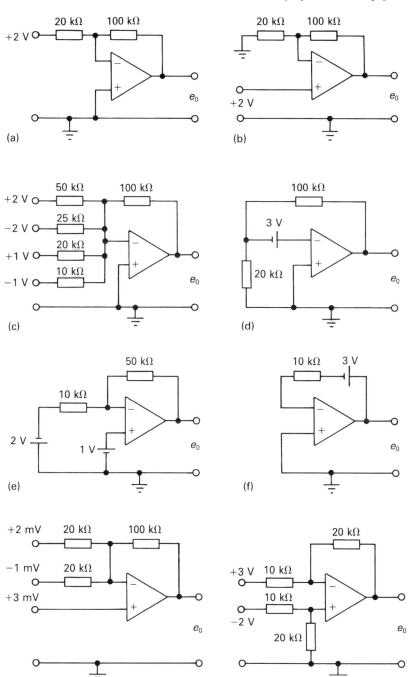

Figure 2.19 *Circuit for Exercise 2.1*

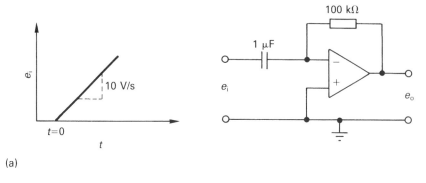

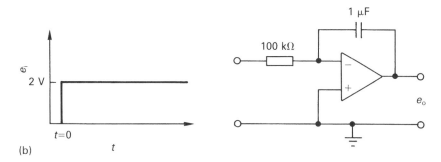

Figure 2.20 *Waveforms and circuits for Exercise 2.2*

difference in the signal gain between the practical and ideal cases.

2.4   A follower type amplifier stage is to be made using an operational amplifier and two external resistors. The circuit used is the one shown in Figure 2.10. The open loop gain is 10^5, $R = 1$ kΩ, $R_f = 5$ kΩ, the differential input resistance is 150 kΩ and the output resistance is negligible. Assuming the common mode input resistance to be infinite, calculate the circuit signal gain and the effective input resistance of the circuit.

2.5   An inverting amplifier is constructed using the circuit shown in Figure 2.11 and where $R = 1$ kΩ and $R_f = 10$ kΩ. Assuming the input source, R_s, to be negligible, calculate the closed loop signal gain for an amplifier open loop gain, A_{OL}, of: (a) 10^6, (b) 10^4, (c) 10^2, (d) 10, and (e) 1. Use your results to plot a graph to show the variation of signal gain against A_{OL}.

2.6   An inverting operational amplifier has an infinite open loop gain, an input resistor of 10 kΩ and a feedback resistor of 100 kΩ. Calculate:

(a) the closed loop signal gain of the circuit; and

    (b) the value of the open loop gain which would make the circuit signal gain only 71% of that calculated in (a).

2.7   An inverting operational amplifier circuit having an input resistor of 10 kΩ produces an output of -6 V for an input of $+1$ V.

    (a) Express the voltage gain in decibels.

    (b) Assuming a signal source of negligible impedance and an ideal operational amplifier, calculate the value of the feedback resistor.

2.8   An inverting operational amplifier circuit has an overall voltage signal gain of 19.5 dB. If the input resistor is 10 kΩ and the feedback resistor 100 kΩ, calculate:

    (a) the feedback fraction, β; and

    (b) the amplifier open loop gain in decibels.

    Assume a negligible signal source impedance.

2.9   The amplifier used with the circuit shown in Figure 2.13 has an open loop gain of 5×10^4 and a differential input resistance of 100 kΩ. $R_f = 1$ kΩ and the input signal is 1 V.

    (a) Calculate the current flowing through R_L .

    (b) If the load resistor were changed from a value of 10 kΩ to 10 Ω calculate the change in load current.

    (c) Calculate the effective amplifier output resistance through which the load current flows.

2.10 The gain of an amplifier is given by the complex expression $A/[1 - j(f_c/f)]$, where A is 200 and the break frequency, f_c, is 10 Hz.

    (a) Draw the Bode approximation diagrams which represent the gain of the amplifier in magnitude and phase.

    (b) For a sinusoidal input signal of 10 mV amplitude, calculate the amplitude of the output signal and its phase relative to the input signal if the signal frequency were: (1) 1 Hz, (2) 10 Hz, and (3) 100 Hz.

3
Basic linear scaling circuits

3.1 Introduction

In Chapter 2 the mechanics of operational amplifiers were discussed in some detail because they are one of the major tools used in analogue signal conditioning circuits. Depending upon the nature of the feedback circuits used, a general purpose operational amplifier usually can be made to undertake any one of a wide range of conditioning applications (Clayton and Newby, 1992: Chap. 4). Subsequent chapters of this book each deal with a different type of application.

This chapter concentrates on a number of signal conditioning circuits which are said to be *linear*. That is, the output follows the input in a linear or straight line relationship of the form:

$$e_o = f(e_i) = me_i + c$$

where m is the slope of the curve representing the relationship between e_o and e_i and c is the value of the output for zero input. This is shown graphically in Figure 3.1.

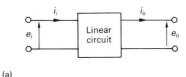

(a)

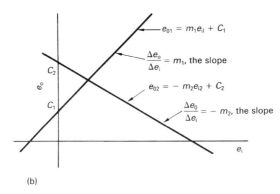

(b)

Figure 3.1 *The linear circuit at (a) has a straight line relationship between its input and output, as shown by the two examples e_{o1} and e_{o2}*

Further, the linear circuit is expected to perform in the same manner at all the frequencies to which it is likely to be subjected; it is independent of frequency. For this to be the case, it follows that the components used in the linear circuits will, in the main, be purely resistive because capacitive and inductive reactances are very frequency dependent.

The linear circuits discussed here will have the purpose of linearly changing, or *scaling*, the signal magnitude or impedance levels, linearly adding or subtracting signals to convert a voltage signal into a current signal, and vice versa. Several of these circuits will use the same, or only slightly modified, operational amplifier configuration. This helps in understanding the operation of the circuit and in its design.

In the following linear applications, the circuits use resistors as the external components which are most frequently connected to the operational amplifier in order to produce a specific signal conditioning function. The circuit designs given do not in general give specific values for resistors and the reader must choose resistor values to suit the required application. This can be a problem in that while the *ratios* of the various resistor values required to achieve a specific scaling factor are known, the actual values to use are not. As a general guideline to the correct selection of resistor values, the underlying principle should be to select values which are not significantly larger than those required to reduce the current drawn from the signal source to an acceptable level. Smaller values of resistor would cause the incoming signal to be distorted due to excessive source loading. Larger resistor values would increase the operational amplifier offset errors due to bias current (see Chapter 2) and when shunted by the inevitable stray capacitance could also limit the operational bandwidth of the circuit. This latter may be a problem for wide band signal conditioning requirements.

General purpose operational amplifiers are designed to supply an output current into a load of no less than 2 kΩ. This output load comprises the true externally connected load in parallel with the feedback resistor connected between the output and input (virtual earth) terminals. Also, in the case of the inverting operational amplifier, the input resistor should be sufficiently large so as to limit the amount of current drawn from the signal source to a level that does not distort the signal.

3.2 Voltage amplification (scaling) and impedance conversion

The circuits in Figure 3.2 are also shown in Chapter 2 but are repeated here for easy reference. These are the basic circuits used for voltage scaling and impedance changing. The great advantage of using these types of operational amplifier circuit is that a very precise function can be performed with a minimum number of precise components being used. For example, the voltage amplification of the circuits in Figures 3.2(a) and 3.2(b) is in each case determined by the value of two resistors, R_1 and R_2. The accuracy of the designed gain is mainly dependent upon the tolerance of these two resistors.

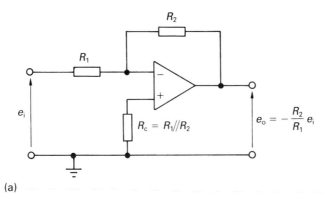

(a)

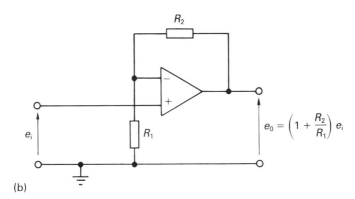

(b)

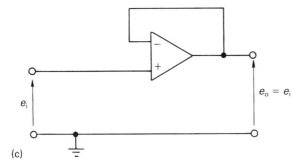

(c)

Figure 3.2 *(a) Inverting amplifier; R_c is added for bias current compensation. (b) Follower with gain. (c) Unity gain follower*

The inverting amplifier in Figure 3.2(a) can have a designed voltage gain from zero up to any higher value, but is limited by the practical requirement that the loop gain must always be sufficient to minimise gain error (see Section 2.3). A further point to remember is the relationship Gain × Bandwidth =

Constant. The higher the gain the less will be the bandwidth. If a very large gain is required, in order to maintain a minimum bandwidth requirement, it may be better to use two lower gain amplifier stages in cascade rather than a single high gain stage.

The inverting and follower circuits both have a relatively low output impedance but their input impedances are very different. The input impedance of the inverting amplifier is largely determined by the finite value of the input resistor, R_1. The non-inverting input of the follower circuit has no external input resistor fitted, yet it presents an extremely high impedance of typically several megaohms. Thus, the follower circuit has the advantage of not loading the signal source yet still produces an appreciable voltage gain. In the case of the unity gain follower (Figure 3.2(c)), while it does not produce a voltage gain, it does show a useful current gain. This makes it an ideal device for changing or matching a high to a low impedance. It is often used as such to buffer (prevent interaction between) a high impedance source (easily overloaded by a heavy current drain) and a low impedance load (demanding a heavy current input). A further advantage of the follower circuits is that since they do not use high value input resistors, which would interact with stray capacitances to produce frequency limitations, they are most suitable for wide bandwidth signal conditioning applications. However, the follower configuration does have the practical disadvantages of being prone to common mode errors and that the input voltage applied to the circuit must not be allowed to exceed the maximum common mode voltage for the amplifier.

The main limitation of the inverting amplifier is that its input impedance is effectively the value of the external input resistor, R_1. Should the application call for a high input impedance to minimise the loading of the signal source, then this demands a large value for R_1. Assuming that closed loop gains greater than unity are required, an even larger value of R_2 must be used. As outlined previously, large value input resistors carrying the small but inevitable amplifier bias currents cause offset voltage error problems. Also, large resistors having even small shunting stray capacitances can cause severe frequency limitations. For example, suppose in Figure 3.2(a) there was a requirement for the input resistance to be 1 MΩ and the signal gain to be 100. This would call for choosing a value of 1 MΩ for R_1 to provide the required input impedance, and so to give the required gain R_2 must be 100 MΩ. Because R_2 is the feedback resistor, R_f, which is shunted by stray capacitances, C_s, then the upper working frequency of the amplifier circuit is given where the gain has fallen to 70.7% of its d.c. value. It has been shown earlier that this frequency is given by the equation $f = 1/(2\pi C_s R_f)$ Hz. Supposing that C_s is 3 pF then the upper working frequency is severely restricted to only 530 Hz. There is also the problem of obtaining such high value resistors that are both accurate and stable.

Because the inverting amplifier configuration with a very high input impedance is often called for, the problem of needing a very high feedback resistor is commonly overcome by the use of a T-resistance network as shown in Figure 3.3. The cost of this circuit, however, is a fall in loop gain and an increase in noise.

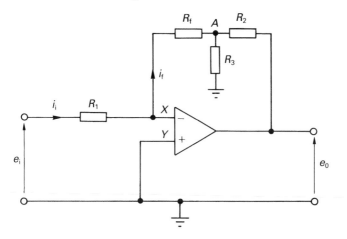

Figure 3.3 *The T-resistance network of R_f, R_2 and R_3 replaces a single high value R_f in the inverting amplifier configuration*

Analysis of the circuit in Figure 3.3 is as follows:

$$i_i = \frac{e_i - e_X}{R_1} = \frac{e_i}{R_1}$$

$$i_f = \frac{e_X - e_A}{R_f} = \frac{e_A}{R_f}$$

Assuming an ideal operational amplifier, no current flows into the inverting terminal and $e_X = e_Y = 0$, so:

$$i_i = i_f$$

and therefore

$$\frac{e_i}{R_1} = -\frac{e_A}{R_f}$$

and so

$$\frac{e_i}{R_1} = -\frac{1}{R_f}\left(\frac{e_o R_3}{R_2 + R_3}\right)$$

Now

$$\text{Gain} = \frac{e_o}{e_i} = -\frac{R_f(R_2 + R_3)}{R_1 R_3}$$

$$= -\frac{R_f R_2 + R_3 R_f}{R_1 R_3}$$

$$\text{Gain} = -\frac{R_f}{R_1}\left(1 + \frac{R_2}{R_3}\right) \tag{3.1}$$

Now for a requirement for $R_1 = 1\,M\Omega$ and a gain of 100 we could choose $R_2 = 200\,k\Omega$ and $R_3 = 10\,k\Omega$. Substituting these values in Equation 3.1 and solving for R_f we find that a value of $5\,M\Omega$ for R_f is sufficient and the previous high value resistor of $100\,M\Omega$ is no longer needed.

3.2.1 Varying the amplification

Very often a variable voltage scaling factor is required. This can be achieved by varying the value of the feedback potentiometer, as shown in Figures 3.4(a) and 3.4(b). The method shown in (a) allows the gain to be varied from zero to a very high value. The disadvantages accompanying this large range of gain are the non-linear mechanical adjustment and the decrease in the input impedance as the gain rises. The method in (b) has a smaller range of adjustment from zero to only R_2/R_1 but the gain changes linearly with mechanical setting of the potentiometer and the input impedance remains constant. A practical point to remember here is that variation of the resistors controlling the gain has two

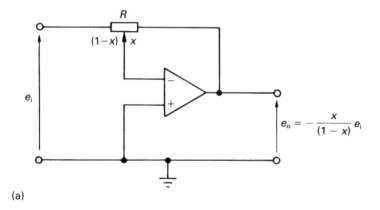

(a)

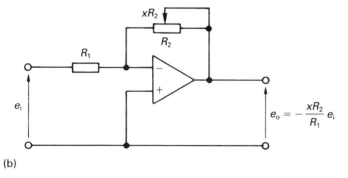

(b)

Figure 3.4 *Variable gain (scale factor) by (a) non-linear control and (b) linear control*

side-effects; the closed loop bandwidth changes as does the offset error caused by amplifier bias currents. In this latter case it can help if a low bias current field-effect transistor (FET) amplifier is used.

3.2.2 Switched amplifier scaling factor

Rather than using a manually adjusted potentiometer, a series of switches can be arranged to replace the input resistor, for example, with different values and so vary the gain in a series of steps. By this method the gain can be switched to any one of an unlimited number of pre-set scaling factors. This can be done manually but it is now common practice to use a programmable integrated circuit chip which is digitally controlled to switch an operational amplifier circuit which uses a high input impedance FET amplifier chip. A typical circuit of a programmable gain operational amplifier is shown in Figure 3.5. A logic 1, (+5 V), on any of the digital inputs closes the appropriate switch in the integrated switch circuit. With a logic 1 on digital input A and logic 0 (0 V) on the other three, SW/A alone closes and the voltage gain of the amplifier is two-fold. A logic 1 on input D alone produces a gain of 100 while a logic 1 on all four inputs results in a gain of 132. With four digital inputs as shown, 16 gain settings or scaling factors can be selected.

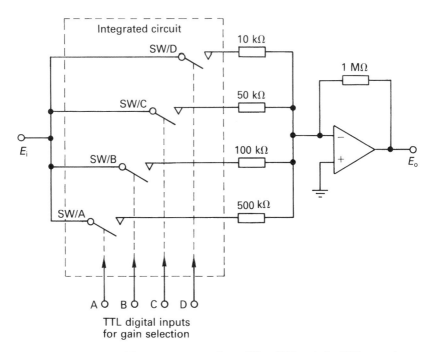

Figure 3.5 *Programmable gain operational amplifier. SW, switch; TTL, transistor–transistor logic*

3.2.3 Voltage controlled scaling factor

This method of controlling the gain of the amplifier is little more than an adaptation of the circuit shown earlier in Figure 3.3. There the effective value of the T-resistance feedback network determines the signal gain, $-R_e/R_1$, and if R_e is varied by voltage adjustment the signal gain will vary in sympathy. Figure 3.6 shows how one manufacturer achieves this by the insertion of a junction gate FET in the T-resistance network. The FET acts as a voltage controlled resistance, r_{ds}, provided it is operated below pinch-off. Assuming that $r_{ds} \ll R_3$, and following a similar mathematical analysis as previously, it can be shown that the effective resistance of the T network is:

$$R_e = R_2 + R_3 + (R_2 R_3 / r_{ds})$$

R_e is a linear function of the controlling voltage V_c, and so the closed loop signal gain, $-R_e/R_1$, also varies linearly with V_c. The amount of variation which can be achieved depends upon the drain to source resistance of the FET. For a practical circuit with component values as shown in Figure 3.6, the maximum gain, when r_{ds} is very small, is approximately 100 and decreases linearly as V_c is made more negative.

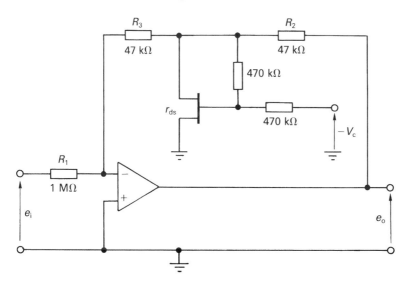

Figure 3.6 *Voltage controlled scaling factor (gain)*

3.3 Voltage summation

It was seen in Chapter 2 that the inverting input terminal of an operational amplifier with negative feedback always assumes virtually the same voltage as is present at the non-inverting terminal. If the non-inverting terminal is earthed then the inverting terminal also becomes a virtual earth. This property enables the inverting operational amplifier to be adapted to form an analogue voltage summing device as shown in Figure 3.7. The parallel inputs, e_1 to e_n, cannot interact with each other because of the isolating effect of the virtual earth on the inverting terminal. In the circuit in Figure 3.7, R_c has been included for bias current compensation purposes, but this is optional. As usual, the input resistor values should only be sufficiently high as to prevent overloading the signal sources. Otherwise, offset error problems caused by even small bias currents flowing through large resistors can arise. If the use of high value resistors is unavoidable, the use of a low input bias current FET operational amplifier should be considered. The ideal summing circuit will permit an unlimited number of parallel inputs, but in practice, in order to maintain an adequate loop gain, the number of inputs is limited. The closed loop gain for the circuit is $1/\beta$ where:

$$1/\beta = 1 + (R_f/\text{Parallel sum of } R_1 \ldots R_n)$$

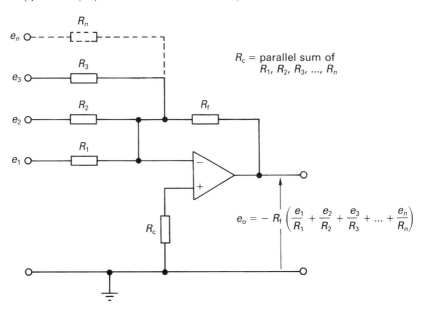

Figure 3.7 *Voltage summing*

3.4 Voltage subtraction

Figure 3.8 will be recognised as a differential input operational amplifier which has an output proportional to the difference of the two input signals, $e_1 - e_2$. The ideal operation of this circuit was covered in Section 2.2.10 where it was assumed that a common signal simultaneously applied to both inputs would result in a zero output. In other words, the circuit should have an infinitely high common mode rejection ratio (CMRR); in practice this is not the case.

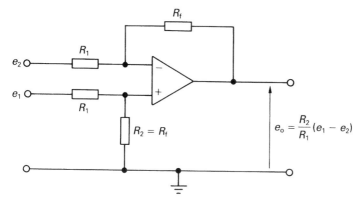

Figure 3.8 *Voltage subtraction*

One reason for this could be a minor mismatch in the values of the two input resistors R_1. The gains of the two signal paths would then be slightly unbalanced and a finite common mode output would result.

The overall CMRR of the above subtraction function is attributable to two sources; the external components and the operational amplifier itself. The CMRR of the complete circuit is obtained from:

$$CMRR = \frac{\text{The differential gain of the circuit}}{\text{The common mode gain of the circuit}}$$

The CMRR is therefore dependent on the possible mismatching of the input resistors within their value tolerances. It can be shown (Clayton and Newby, 1992: Appendix A3) that the CMRR of the complete circuit is more attributable to the use of mismatched resistors, both within the same tolerance, than to the operational amplifier employed. The overall CMRR of the circuit caused by both effects is:

$$\text{Total CMRR} = \frac{\text{Resistor CMRR} \times \text{Op-amp CMRR}}{\text{Resistor CMRR} \pm \text{Op-amp CMRR}}$$

The above expression shows that the two effects may be series aiding or opposing so that the total CMRR of the circuit may be greater or smaller than that of the operational amplifier used. It is possible purposely to trim the value of one of the input resistors, within its tolerance, such that the resistor CMRR is equal and opposite to that of the operational amplifier. In practice the total CMRR can be improved by a factor of 100 but since resistor values vary with temperature the high CMRR thus obtained is not stable.

As ever, there are applications for these differential input operational amplifiers where both the differential and the common mode input impedances must be large to prevent loading the signal sources. In Figure 3.8, the differential input resistance is $R_1 + R_1 = 2R_2$ while the common mode input resistance at both inputs is effectively $R_1 + R_2$. If these resistors are simply given large values we encounter the usual problems associated with the stray capacitances having greater effect, especially at high frequencies, and a lowering of the CMRR results. The high value resistors also increase the offset errors caused by the flow of the inevitable bias currents.

While the simple circuit in Figure 3.8 is often used in non-critical applications, in order to give the high input impedance and high noise rejection required by many instrumentation applications, two or more operational amplifiers are used. The resulting high performance amplifier is often referred to as an 'instrumentation amplifier' and can be obtained commercially in modular or integrated circuit packages.

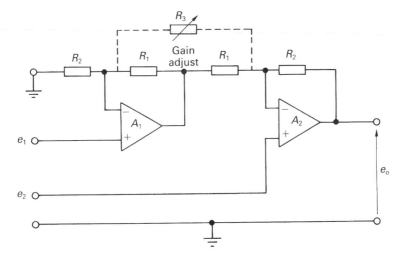

Figure 3.9 *Two followers connected as a high input impedance differential amplifier*

Figure 3.9 shows how two followers can be connected in a differential input configuration to produce the required enhanced performance. Assuming ideal operational amplifiers, the gain of A_1 has a gain of $1 + R_1/R_2$ and produces an output voltage of $e_1(1 + R_1/R_2)$. Amplifier A_2 has two inputs, the output of A_1, which experiences a gain of $-R_2/R_1$, and e_2, which experiences a gain of $1 + R_2/R_1$. The final output, e_o, is the algebraic sum of the two individual amplifier outputs and is given by:

$$e_o = (e_2 - e_1)(1 + R_2/R_1)$$

If gain adjustment is required, this can be achieved by fitting a single variable resistor, R_3. This affects the gain of A_2 because it is now determined by R_2 and the parallel combination of R_1 and R_3.

Now the output is given by:

$$e_o = (e_2 - e_1)[1 + (R_2/R_1) + 2R_2/R_3]$$

Note that when $e_1 = e_2$ in magnitude and phase, as would be the case with an unwanted noise voltage appearing at both the differential inputs, then the noise is not amplified but cancelled.

A practical circuit comprising three OP-27 low noise operational amplifiers is shown at Figure 3.10. This circuit is especially useful for instrumentation and professional quality audio systems.

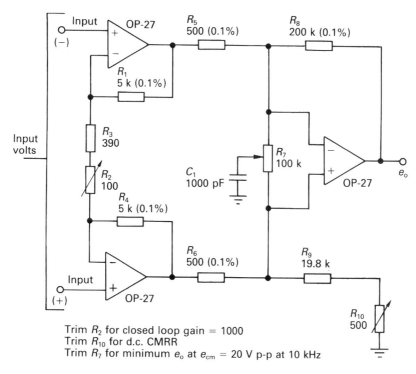

Figure 3.10 *A practical instrumentation quality amplifier using three operational amplifier integrated circuits. p-p, peak to peak*

3.5 Current amplification (scaling)

When stimulated, some types of transducer, light sensors for example, produce an output current change rather than a voltage change. The circuits discussed thus far are designed specifically for voltage inputs. The circuits described in this section are for use with systems which have a current input.

3.5.1 Current-to-voltage conversion

It was shown in Chapter 2 that the ideal operational amplifier can act as a current-to-voltage converter. The simple circuit is shown in Figure 3.11.

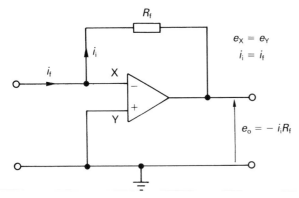

Figure 3.11 *The ideal operational amplifier converts* i_i *to* e_o

Because no current flows into the ideal operational amplifier input terminals and the circuit feedback action maintains the voltage at X at the same value as that at Y (zero in this case), the input current i_i all flows through the feedback resistor R_f. The voltage output is therefore $e_o = i_i R_f$.

The more practical situation is depicted in Figure 3.12 where the current source is shown as comprising a constant current generator shunted by R_s and C_s in parallel. Gain stability can be a problem at some frequencies and this is usually taken care of by the fitting of the feedback resistor R_f. A basic cause of the instability is the capacitance, $C_{s,}$ of the current source. It causes a phase lag in the signal feedback to the amplifier inverting terminal and this can lead to

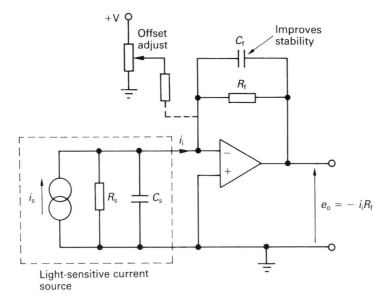

Figure 3.12 *Current-to-voltage converter – practical considerations*

insufficient phase margin at the higher frequencies. The introduction of R_f produces a compensating phase lead.

Initial offset error can be removed by the injection of a small compensating voltage into the inverting input terminal as shown in Figure 3.12. The temperature drift of the amplifier input bias current is then the limiting factor determining the current-to-voltage conversion accuracy. For this reason, when converting small currents it is useful to use a FET type amplifier.

3.5.2 Current summation

The basic circuit arrangement shown in Figure 3.13 can be used to add currents from separate signal sources. In order to ensure gain stability by maintaining an adequate phase margin, the feedback capacitor, C_f, is now determined by the sum of the individual source capacitances now comprising C_s.

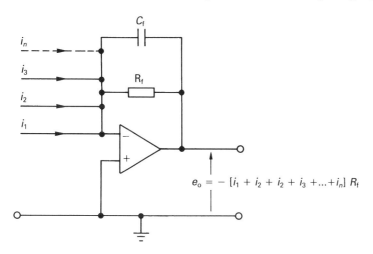

Figure 3.13 *Current sum to voltage conversion*

3.5.3 Current difference to voltage conversion

The circuit in Figure 3.14 shows how two operational amplifiers can be connected to form a current difference measuring function without introducing a voltage drop into the circuit producing the current. The zero volt drop condition is achieved by the current to be measured being passed to the virtually earthed inverting terminal of an operational amplifier which has its non-inverting terminal grounded. Amplifier A_1 acts as a current inverter by producing $-i_2$ and amplifier A_2 is a current-to-voltage converter and produces an output proportional to $i_1 + (-i_2)$. Examination of amplifier A_1 shows that when equal resistors are connected between its inverting and non-inverting terminals, equal currents must flow through these two resistors in order to

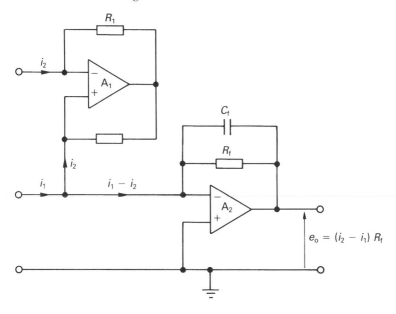

Figure 3.14 *Conversion of current difference to voltage*

maintain the two terminals at the same potential. The input current to amplifier A_2 is thus $i_1 - i_2$ and its output voltage is $e_o = -(i_1 - i_2)R_f = (i_2 - i_1)R_f$.

If it is permissible to introduce a small voltage change into the sensing circuit, the more simple current differencing circuit shown in Figure 3.15 may be used. The simplicity is obtained for the cost of introducing an i_2R voltage drop into the sensing circuit.

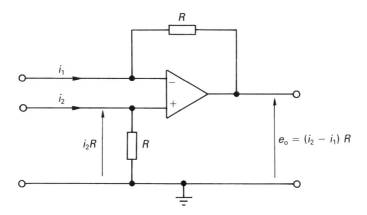

Figure 3.15 *Simple circuit for current difference to voltage conversion. i_2R is the voltage introduced into the circuit under test*

3.6 Voltage-to-current conversion

Several signal conditioning functions require a constant current source, the output of which can be controlled by the adjustment of a voltage. Such applications include resistance measurement and testing and the control of coil currents in the production of magnetic fields. There is a wide variety of operational amplifier circuits which may be employed for constant current working; the nature of the load conditions dictating the one most suitable. A selection of example circuits follows in the next sections.

3.6.1 Voltage-to-current conversion – floating load

The basic inverting and non-inverting operational amplifier circuits are the simplest to design because all that is necessary is to use the load as the feedback impedance and the current flowing is governed by the input voltage, e_1, and the input resistor, R_1. Figure 3.16 depicts the two cases. The inverting amplifier case requires the signal source to provide the load current. This is not the case with the non-inverting amplifier where the load current is provided by the amplifier output. However, care should be taken to ensure that the amplifier output voltage required to drive the necessary current does not exceed the amplifier voltage output rating.

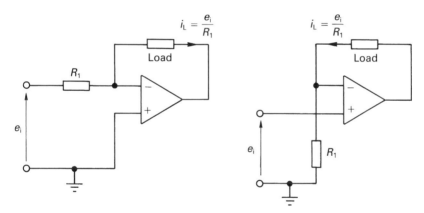

Figure 3.16 *Simple voltage-to-current converter (floating load)*

3.6.2 Voltage-to-current conversion – earthed load

The circuit configuration is shown in Figure 3.17 where the load impedance is earthed. The normal operational amplifier action is such that through the feedback circuit the two amplifier inputs are driven to virtually the same potential. This causes a voltage approximately equal to e_i to appear across the resistance R through which a current of e_i/R flows. For practical purposes, all this current can be regarded as also flowing through the load

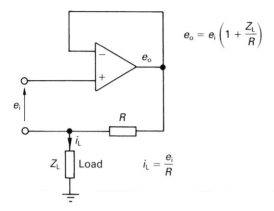

$$e_o = e_i \left(1 + \frac{Z_L}{R} \right)$$

$$i_L = \frac{e_i}{R}$$

Figure 3.17 *Voltage-to-current converter (floating signal source; earthed load)*

impedance, Z_L, since virtually no current can pass into the high impedance non-inverting input of the amplifier. Again, care should be taken to ensure that the voltage limitations of the operational amplifier are not exceeded.

3.6.3 Current sources and sinks

A simple means of obtaining a constant load current which is voltage controlled is to use a resistor, a transistor and an operational amplifier as shown in Figure 3.18. A limitation of this method is that the current flow through the load is restricted to one direction only. The controlling voltage, e_i, is fed into

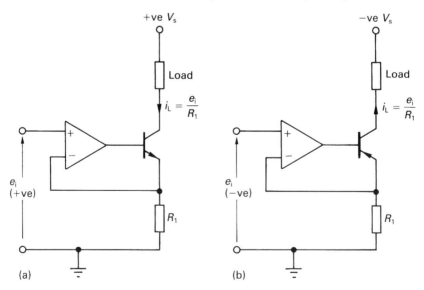

Figure 3.18 *(a) Current sink; (b) current source*

the non-inverting input of the operational amplifier. By virtue of the feedback circuit, the inverting input also assumes e_i and hence the transistor emitter current, flowing through R_1, is given by e_i/R_1. For practical purposes the transistor emitter and collector currents are the same and are controlled by the very small base current provided by the operational amplifier output. If an npn transistor is used, as in Figure 3.18(a), the load current 'sinks' into the controlling circuit whereas the npn transistor configuration in Figure 3.18(b) is said to 'source' the load current. An advantage of using these circuits is that the load currents they handle can be well in excess of the output current rating of the operational amplifier. The operational amplifier merely provides the small transistor base current and the load current is limited only by the level of the transistor saturation current.

3.7 A.c. amplifiers

Even though operational amplifiers are mainly regarded as very high gain d.c. devices they are also used to great effect in a.c. circuits. D.c. effects are prevented from entering the input by the addition of a blocking capacitor. However, it is important to provide a d.c. bias path to the two amplifier inputs.

3.7.1 Inverting a.c. amplifier

The circuit used for this function is shown in Figure 3.19. C_1 acts as the d.c. blocking capacitor and in conjunction with the input resistor R_1 determines the lower frequency -3 dB point as being $1/(2\pi C_1 R_1)$. The upper frequency limit will be determined by the upper frequency gain roll-off of the operational amplifier. Assuming that the capacitor C_1 has negligible reactance in the operating frequency band, the gain of the circuit will be given by R_2/R_1. The d.c. bias current for the inverting terminal is obtained through R_2.

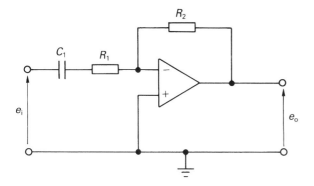

Figure 3.19 *Phase inverting a.c. amplifier*

3.7.2 Non-inverting a.c. amplifier

This is depicted in Figure 3.20. The capacitors act as d.c. blocks and the addition of R_3 is necessary to provide a d.c. path to earth for the non-inverting input. R_3 also decides the circuit input resistance. The lower frequency gain response curve will show two distinct breaks; one at a frequency given by $1/(2\pi C_1 R_1)$; the other by $1/(2\pi C_2 R_3)$. The circuit gain is that of the typical follower, $1 + (R_2/R_1)$.

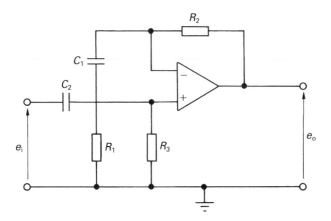

Figure 3.20 *Non-inverting a.c. amplifier*

3.7.3 High input impedance amplifier

The addition of the d.c. bias path resistor R_3 in the previous follower circuit reduces the normally high input impedance of the standard follower circuit. A high input impedance can be restored to the circuit by the technique known as *bootstrapping*. The idea is to apply the input a.c. signal to each end of the resistor R_3 such that there is little effective signal voltage across R_3 and hence little signal current flows through it (Figure 3.21). R_3 therefore appears to have an enhanced resistance and raises the input impedance accordingly. The bootstrapping effect is obtained by the positive feedback of the input signal to the lower end of R_3 through R_2 and C_1. The effective increase in the resistance value is by the factor equal to the loop gain.

Exercises

3.1 For Figure 3.3 show that the gain of the circuit is approximately given by: $-(R_f/R_1)[1 + R_2/R_3]$.

3.2 If in Figure 3.5 a logic 1 transistor–transistor logic (TTL) input closes the relevant switch, calculate the output voltage for the case where $E_i = 2$ mV and the TTL digital input word is (a) 1001 and (b) 0110.

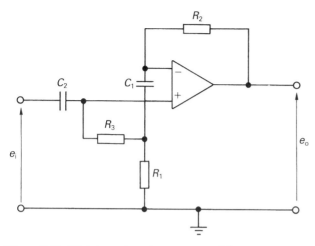

Figure 3.21 *High input impedance a.c. amplifier*

3.3 For Figure 3.8 verify the relationship, $e_o = (R_2/R_1)[e_1 - e_2]$.

3.4 Suppose for Figure 3 8 that initially $R_1 = 10$ kΩ and that $R_2 = R_f = 50$ kΩ.

    (a) Calculate the output voltage if $e_1 = 2$ V and $e_2 = 1$ V.

    (b) Now calculate the percentage error in the output volts if the resistor R_2 became defective and doubled in value.

3.5 With reference to Figure 3.13 show that the output voltage is proportional to the sum of the input currents and explain the purpose of the capacitor C_f.

Reference

Clayton, G. and Newby, B. (1992) *Operational Amplifiers*, 3rd edn, Chap. 4, Butterworth-Heinemann, Oxford.

4
Non-linear circuits

In the previous chapter the feedback circuits associated with the different operational amplifier configurations used pure resistors; inductances and capacitances were generally not required. This caused the input and output voltage relationships for the different functions to be linear and, in the main, independent of the input signal frequency. If frequency conscious reactive components are used in the feedback circuits then the output can be expected to respond in a non-linear manner.

This non-linear response has many uses in the field of signal conditioning. The output can be designed to relate to the input in a multiplicity of useful ways. For example, the output may be the mathematical square of the input; its square-root; a trigonometrical function of it; or perhaps its natural logarithm. In this latter respect, the logarithm of different functions can be produced and then processed further to result in mathematical operations such as multiplication, division or the evaluation of the powers or roots of numbers.

Another use for these non-linear function generators is to simulate a specific response for a particular input. This modelling of a real life situation can help in predicting the outcome of a design change without having to try the real thing. This can save time, trouble and expense. The following sections will examine a range of the techniques used to implement different non-linear signal conditioning functions.

4.1 Specific non-linear amplification

Suppose that we wish to produce a non-linear relationship between the current flowing through a circuit for an applied linear input voltage. This could be achieved using a non-linear circuit element, as opposed to a pure resistance, in the circuit shown in Figure 4.1. The circuit current flowing would be a non-linear function of the input voltage. This non-linear current could be applied to the inverting input terminal of an operational amplifier as shown by Figure 4.2. With no current flowing into the ideal amplifier terminal, all the non-linear current would be routed through the feedback resistor, R_f, and consequently the output voltage, e_o, would be the required non-linear function of the input voltage, namely, $e_o = -R_f f(e_i)$ where $f(e_i)$ is a non-linear function of e_i.

Figure 4.3 shows the non-linear element connected in the alternative position as the feedback resistor leaving the input resistor, R, as a normal linear device. The input current, once again, is virtually the same as that in the feedback circuit and so we can equate the expressions for the two:

$$i_f = -f(e_o) = i_i = e_i/R$$

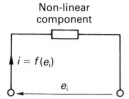

Figure 4.1 *The current in the circuit is in a non-linear relationship with the input voltage*

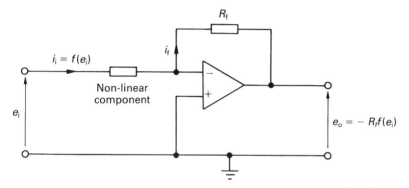

Figure 4.2 *Operational amplifier circuit producing a non-linear transfer function*

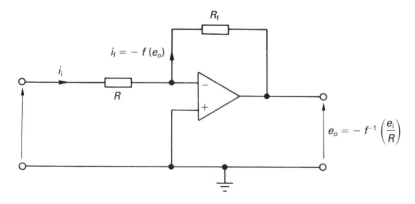

Figure 4.3 *Operational amplifier producing a non-linear inverse function.* R_f *is the non-linear component*

and therefore,

$$e_o = -f^{-1}(e_i/R)$$

where f^{-1} denotes the inverse function of e_i/R. (*Note*: the term *inverse function* simply means reversing the process as, for example, mathematical integration is the inverse of differentiation.)

Two of the major difficulties associated with practical non-linear amplifiers is getting them to perform satisfactorily over a wide current range and at the same time be insensitive to changes of temperature. In general there are two methods of approach to these problems. One is to produce an approximation of the non-linear response curve by a series of connected short straight lines; the other is to use a component which has the necessary non-linear qualities. The first method is generally known as one of *synthesis* and is examined next.

4.2 Synthesised non-linear responses

Figure 4.4 illustrates the principle of synthesising a smooth curve by a series of short straight lines. The shorter the straight lines the more accurate is the synthesis but the more complicated and expensive is the circuitry required.

Figure 4.5 shows a well-known method of synthesising a non-linear function using a network of biased diodes and resistors together with a current summing operational amplifier. In the absence of e_i, $-e_{ref}$ is applied through R_2 to the anode of ideal diode D_1 which is thus reverse biased and does not conduct. The progressive application of a positive voltage for e_i now raises the potential at the anode of D_1 and when it reaches that of its cathode (virtual earth) D_1 starts to conduct. The current which flows through D_1 is i_1, all of which flows through the feedback resistor, R, of the ideal operational amplifier to give an output voltage $e_o = -Ri_1$. The values of the resistor complex $R_1, R_2, \ldots, R_{10}$ should be chosen so as to make the remaining diodes start conducting in progressive order, at the various breakpoint voltages, as e_i is increased in value. Eventually all the diodes will be made to conduct and the current flowing through R will be the sum of the five diode currents, so producing $e_o = i_T R$. As a general guide, for most practical applications, five to eight linear segments are adequate to approximate to functions requiring about a 90° curve, for example for a square law where $e_o = (e_i)^2$.

Should it be required to work with negative inputs for e_i, then this can be catered for by using additional input circuits with the diode and reference voltage polarities reversed.

The above explanation of the functioning of the circuit shown in Figure 4.5 assumes that there is no voltage drop across the diodes and that the circuit works equally well at all temperatures. Neither assumption is true and so should the simple synthesiser circuit be unsuitable in either of these respects then the appropriate compensating circuitry must be added. This clearly adds to the complication and expense of providing the required function and it is at this juncture that the purchase of a suitable commercially available module or integrated circuit should be considered.

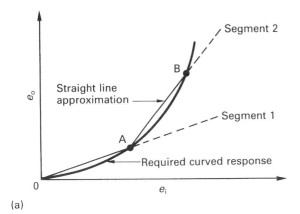

(a)

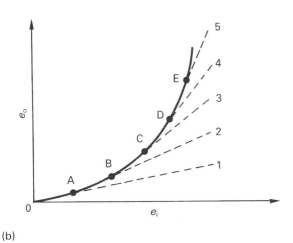

(b)

Figure 4.4 *Straight line approximation of a non-linear function (synthesis). (a) Two segments give only a rough approximation to the time curve. (b) Five segments give a better approximation to the true curve*

4.3 Logarithmic response using a non-linear component

Figure 4.6 shows how a non-linear element, a diode, can be connected into the feedback circuit of an operational amplifier to produce a logarithmic conversion of the input voltage signal. This is possible because the current flow through a diode and the voltage across the junction have a natural log relationship. The *logging* action of the diode is explained by examination of Shockley's equation (sometimes called the *ideal diode equation*) for a single pn junction.

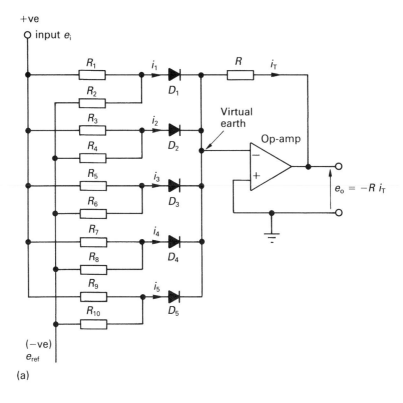

(a)

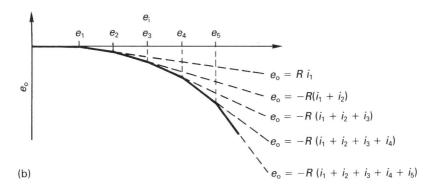

(b)

Figure 4.5 *Synthesised non-linear response comprising five linear segments. (a) Ideal diodes and operational amplifier are assumed. Maximum*
$i_T = i_1 + i_2 + i_3 + i_4 + i_5$. $e_i = e_1$ when D_1 starts conducting, $e_i = e_2$ when D_2 also conducts, etc. (b) $e_1, e_2 \ldots, e_5$ are 'break voltages'

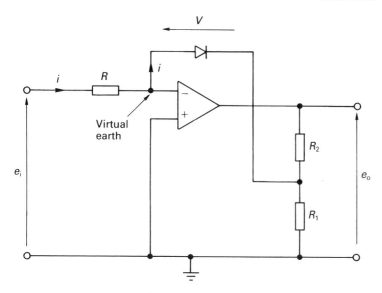

Figure 4.6 *Logarithmic amplifier using a diode as the non-linear log element*

Shockley's equation states:

$$I = I_o \left(\exp \frac{qV}{kT} - 1 \right)$$

where I is the current through the junction (A), I_o is the theoretical reverse saturation current (A), V is the voltage across the junction, q is the magnitude of the electron charge (1.6×10^{-19} C), k is Boltzmann's constant (1.38×10^{-23} J/K), and T is the temperature (K).

If we put the values of the constants into the equation we find that kT/q is approximately 26 mV at 27°C. So with a voltage across the pn junction of about 100 mV, the exponential term is much larger than unity and the equation for the forward biased junction current can be written as:

$$I \approx I_o \exp \frac{qV}{kT} \quad \text{(where } V > 100 \, \text{mV)}$$

We can rearrange this equation and take logarithms of both sides and obtain:

$$\log_e \frac{I}{I_o} = \frac{qV}{kT}$$

or, rearranging,

$$V = 2.3 \frac{kT}{q} \log_{10} \frac{I}{I_o} \tag{4.1}$$

If we now plot a graph of $\log_{10} I$ against V a straight line through zero results and it has a slope of $2.3kT/q$ V/decade of current change. (It is of interest to note that this means that at a temperature of 27°C, changing the junction voltage by approximately 60 mV will result in a 10-fold increase of junction current flow.)

Applying Equation 4.1 to the circuit in Figure 4.6 we can say that because the amplifier inverting input terminal is virtually at earth potential, then

$$V = \frac{R_1}{R_1 + R_2} e_o$$

but from Equation 4.1

$$V = 2.3 \frac{kT}{q} \log_{10} \frac{I}{I_o}$$

Equating the two expressions and rearranging, we have

$$e_o = -\frac{R_1}{R_1 + R_2} 2.3 \frac{kT}{q} \log_{10} \frac{I}{I_o} \qquad (4.2)$$

In the circuit shown in Figure 4.6, $i = e_i/R$ is the value of I appearing in Equation 4.2. Also, the values of R_1 and R_2 can be chosen so as to produce a convenient scaling factor rather than have the 60 mV/decade of current change mentioned earlier; 1 V/decade of current change is usually preferred.

Because of the diode's temperature dependence, the simple logging circuit in Figure 4.6 has a limited performance even if it is used under strictly controlled ambient conditions. The basic ideal diode equation is itself dependent upon the operating temperature, T, and the voltage output, e_o, evidently increases linearly with temperature. This can be partly compensated for (see Equation 4.2) by using a temperature sensitive resistor for R_1 which increases its value with increases of T.

The temperature dependence of I_o has an even more marked effect on the logging conversion accuracy. Not only does I_o approximately double for every 10°C rise in temperature, but it also changes non-linearly. The use of matched diodes in two separate logging circuits used in conjunction with a third circuit operating in a subtracting role can compensate for this by eliminating the term I_o.

Finally, and most importantly, is the fact that practical diodes do not generally accurately perform as predicted by Equation 4.1. It has been said that the junction current is not governed by a single diffusion process but several. Further, at the higher junction currents, logging accuracy is further eroded because the resistance of the basic bulk semiconductor material causes a voltage drop so that the junction voltage is less than that applied across the whole diode.

The general problems caused by the temperature dependence of the diode makes them unsuitable for accurate logging conversions unless the temperature is controlled and the current change is limited to two or three decades. The transistor is a more convenient logging device and we shall now consider how it can be used in circuits to obviate the shortcomings of the diode.

The bipolar transistor comprises two pn junctions and therefore the mathematical equations which model the flow of currents around the device are somewhat more complicated than those for the simple diode. It is sufficient for this book to say that work in this field has shown that, provided the collector to base voltage of a common base connected transistor is made equal to zero, then the collector current is of the form:

$$-V_{EB} = 2.3 \frac{kT}{q} \log \frac{I_C}{I_o} \tag{4.3}$$

The requirement of ensuring that $V_C = 0$ is met by connecting the transistor in the feedback loop of an inverting operational amplifier having its non-inverting terminal earthed. The virtual earth produced at the inverting terminal effectively makes $V_C = 0$. The circuit is shown in Figure 4.7; it is known as the *transdiode* logging configuration. The output voltage is effectively taken from the emitter so that $e_o = V_{EB}$. An alternative logging circuit is obtained by connecting the feedback transistor as a diode by shorting its collector and base together; the circuit shown in Figure 4.8 is the result. Of the two circuits, the transdiode connection can cope with the widest range of input currents for accurate logging purposes. Its current range is of the order of 10 decades with its higher current handling capability being limited to about 10 mA by the effects of the basic semiconductor bulk resistance. However, the disadvantages of the earthed base transdiode connection are its incapability of accepting negative inputs (unless the transistor is replaced by a pnp type) and the closed loop frequency stability problems introduced by the use of a feedback loop having a frequency dependent gain.

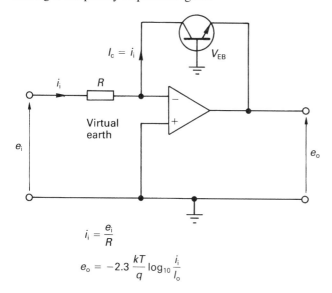

Figure 4.7 *Transdiode logging amplifier*

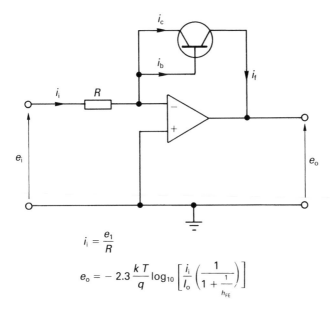

$$i_i = \frac{e_1}{R}$$

$$e_o = -2.3 \frac{kT}{q} \log_{10} \left[\frac{i_i}{I_o} \left(\frac{1}{1 + \frac{1}{h_{FE}}} \right) \right]$$

Figure 4.8 *Log amplifier using a diode connected transistor*

The alternative diode connection has a smaller logging range but, since it has no gain capability with its collector and base shorted together, it does not have the same frequency stability problems. Further, being effectively a two terminal device, it can operate successfully with a reversing polarity input signal. Also, the feedback current is not that due to i_c alone but i_c and i_b; i_c and i_b are related by the transistor forward d.c. current gain, h_{FE}. Th, $i_f = i_c + i_b = i_c(1 + 1/h_{FE})$ and this modifies Equation 4.3 to read:

$$-e_o = -V_{BE} = 2.3 \frac{kT}{q} \log_{10} \left[\frac{i_c}{I_o(1 + 1/h_{FE})} \right] \tag{4.4}$$

Examination of Equation 4.4 shows that in order to keep the error term $(1 + 1/h_{FE})$ as near unity as possible, the d.c. current gain of the transistor used should be as high as possible. The logging characteristic of a typical diode connected transistor is shown in Figure 4.9. Note that the logging range is confined to values of V_{EB} which are less than the normal 0.6 V taken as the typical value for a silicon transistor to start its normal useful conduction.

Figure 4.10 shows a third alternative logging circuit which is useful for applications that require a minimum current loading to be placed on the operational amplifier. In this case the operational amplifier is required to supply only the base current while the feedback loop current to the inverting terminal is the transistor emitter current. As with the previous logging circuit configurations, a condition for accuracy over a reasonable range is to keep the

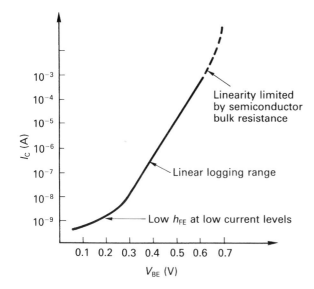

Figure 4.9 *Typical logging characteristic of a diode connected transistor*

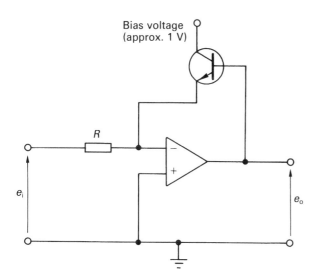

Figure 4.10 *Transistor log amplifier*

operational amplifier summing point as near to zero as possible and this is achieved in Figure 4.10 by applying a bias voltage of about 1 V to the feedback transistor. The disadvantages of this circuit are the lack of signal reversibility and having to provide a separate bias voltage supply.

4.4 Practical logarithmic amplifiers

While the previous sections of this chapter have discussed the basic theory of logarithmic amplifiers, the modern way of producing a particular function requiring a logarithmic conversion of a signal voltage or current is to use a commercially available integrated circuit. Typical of these integrated circuits is the TL441AM. In order that the reader may have a first hand experience of the information and application details which manufacturers now make available, the complete Texas Instrument data sheet for the TL441AM logarithmic amplifier integrated circuit is reproduced in Appendix 2 to this book.

Because logarithmic amplifiers make prolific use of operational amplifiers, the inadequacies of these are still present and need to be catered for. The usual problems are summarised below but are discussed in greater detail in Clayton and Newby (1992).

4.4.1 Closed loop stability

In order that an amplifier shall remain stable (non-oscillatory) it is required that the feedback loop gain shall be less than unity at which the phase shift around the loop is 180°. In addition to the usual capacitive effects which can cause instability in any feedback amplifier, the logging amplifier has the additional problem of having a transistor in the feedback loop which introduces a pronounced non-linear gain factor; the feedback is greater and therefore the gain is smaller at the higher input currents. This problem can be rectified by the judicious use of compensating CR networks.

4.4.2 Balancing offset errors

In many cases the lower limit to the useful logging range is determined not so much by the performance of the transistor in the feedback loop but by the basic amplifier offsets. In order to reduce the offset errors to a minimum and to maximise the logging range, the voltage and current offsets can be separately eliminated by compensating biasing techniques. These are often recommended by the amplifier manufacturer and should be adopted whenever possible.

4.4.3 Reverse voltage protection

Even though the input signal to the logging amplifier may be only small, should it suffer a polarity reversal damagingly large emitter bias voltages can be generated. For this reason it is advisable to provide diode protection circuits if the manufacturer of the amplifier has not included them.

4.4.4 Temperature compensation

As mentioned in Section 4.3, the logging amplifier with its non-linear semi-conductor active element in its feedback circuit is extremely susceptible to temperature changes. Of all the temperature change effects, perhaps the temperature dependence of the reverse saturation leakage current, I_o, is the most important. While a strictly controlled ambient temperature will help, the usual technique adopted is to use a pair of matched logging transistors having the same values for I_o. The two voltage outputs thus obtained for a common input are then differenced, which cancels the effect of I_o. The voltage difference is thus independent of I_o and is proportional to the logarithm of the ratio of the two transistors' collector currents, one of which is usually made into an adjustable (and controllable) fixed reference.

4.5 Practical applications

Appendix 2 contains a large amount of technical detail concerning the construction and operation of the commercially available multi-stage TL441AM monolithic logarithmic amplifier together with notes for its use in a range of practical applications. These include:

- Output slope and original adjustment.
- Utilisation of the separate stages.
- Utilisation of parallel inputs.
- Logarithmic amplifier with an input voltage range greater than 80 dB.
- Multiplication or division.
- Raising a variable number to a fixed power.
- Raising a fixed number to a variable power.
- Dual-channel radiofrequency logarithmic amplifier with a 50 dB input range per channel at 10 MHz.

Exercises

4.1 For Figure 4.11 show that $e_o = -R(e_i/R_a + E_{ref}/R_b)$.

4.2 Figure 4.12 shows a circuit for an n segment synthesis of a non-linear response. Assume the diodes and the operational amplifier to be ideal.

(a) Deduce an expression for the current flowing in the feedback loop and from this show that the output voltage is

$$e_o = e_i R(1/R_{a1} + 1/R_{a2} + \ldots + 1/R_{an}) - E_{ref} R(1/R_{b1} + 1/R_{b2} + \ldots + 1/R_{bn})$$

(b) Sketch the graph of the relationship between e_o and e_i as e_i is increased from zero. Show clearly the 'break voltages'. (Hint: see Figure 4.5.)

(c) Show that the nth break voltage occurs when e_i has risen to $E_{ref}(R_{an}/R_{bn})$.

4.3 Calculate the output voltage of the logarithmic amplifier circuit shown in Figure 4.8 if the input current is 10^{-9} A and the ambient temperature is 27°C. Assume that $I_o = 10^{-12}$ and that $h_{FE} = 100$. What does the output voltage become if the temperature (a) rises to 37°C or (b) falls to 17°C? (Hint: remember that temperature affects the magnitude of I_o.)

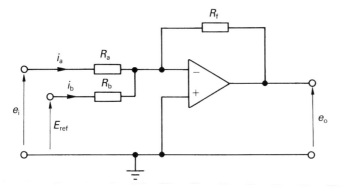

Figure 4.11 *Circuit for Exercise 4.1*

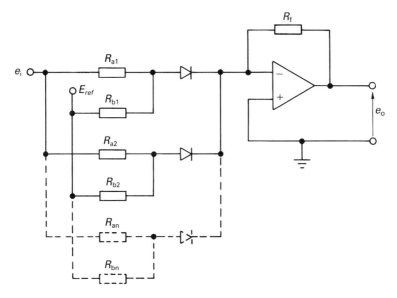

Figure 4.12 *Circuit for Exercise 4.2*

Reference

Clayton, G. and Newby, B. (1992) *Operational Amplifiers*, 3rd edn, Chap. 5, Butterworth-Heinemann, Oxford.

5
Integrators and differentiators

This chapter is concerned with the circuitry required to perform the mathematical functions of integration and differentiation together with their practical applications in instrumentation and process control. Integration will be considered first. The operation of the ideal integrator will be examined and this will be followed by discussion of how the practical circuit deviates from this ideal and how corrective action or allowances can be made. Examples will be given of how the practical integrator can be used to generate positive-going or negative-going ramp voltages, measure electrical charge, produce triangular or sawtooth waveforms, perform timing functions and the like.

5.1 The ideal integrator

The most common circuits used for providing a precise integration function employ an operational amplifier. It will be recalled that the two *golden rules* concerning the working of an ideal operational amplifier are:

(a) the input impedance to the amplifier is infinitely high so that no current flows into the amplifier input terminals, and
(b) the action of the amplifier is always to make the output voltage assume such a value that by means of the feedback circuit the two amplifier input terminals are at the same potential.

Rule (a) means that in the case of the inverting operational amplifier connection the whole of the input signal current must flow in the feedback circuit whatever components it may contain. Rule (b) implies that if the non-inverting terminal is connected to earth then the inverting terminal is also at earth potential and prevents the accumulation of any charge at that point.

In addition to the operational amplifier the integrator requires a capacitor in its feedback circuit. There are two important relationships concerning capacitors, namely

$$Q = VC \quad \text{and} \quad Q = It$$

where Q is the capacitor charge (C), V is the voltage across the capacitor, C is the capacitance (F), I is the steady current flowing into the capacitor, and t is the time for which current I flows.

But the current does not normally flow steadily into a capacitor because of the voltage build-up across the capacitor as it accumulates charge. This decay of charging current is expressed mathematically by considering the instantaneous current, i, and summing this over the interval for which it is flowing.

So, based on the two expressions above we can say

$$VC = It$$

and substituting instantaneous values and integrating over the charge time interval this becomes

$$v = \frac{1}{C} \int i \, dt$$

We can now combine the above principles to a standard inverting operational amplifier connection as shown in Figure 5.1(a). The input current $i_i = e_i/R$ flows through R to the ideal amplifier summing point. No current enters the inverting terminal and so all i_i flows into the feedback capacitor encouraged so to do by the amplifier negative output voltage. The flow of charging current into the capacitor is maintained at the precise value necessary to keep the inverting input terminal at a virtual earth potential. The amplifier output voltage is therefore the same as that across the capacitor:

$$e_o = -V = -\frac{1}{C} \int i_i \, dt$$

$$e_o = -\frac{1}{CR} \int e_i \, dt$$

The input impedance is equal to the resistance R while the output impedance is of a low value because of the inherent negative feedback through C. The *characteristic time* of the circuit is given by the product CR. The value of $1/CR$ decides the slope of the ramp output created by a step input. Suppose the value of C were 1 µF and that of R were 1 MΩ, then an input step voltage of $+1$ V would cause a current of 1 µA to flow into the capacitor C. In order to maintain this capacitor charging current at a steady value the amplifier output voltage would need to fall at the linear rate of -1 V/s. The conversion of the step input into a ramp output (a $+1$ V step input would cause a positive-going ramp output) constitutes the integrating action of the circuit. Figure 5.1(b) shows the waveform of a $+1$ V step input and Figure 5.1(c) the corresponding integrated output response. Figure 5.1(b) also shows the practical effects of the capacitor leakage resistance and the limitation on the output voltage magnitude set by the supply voltage V_s.

It is of interest to note that should the input charging current to the capacitor be cut off in midflow, then the capacitor would remain (hold) at its instantaneous charged state. A practical requirement would be the ability to reset the capacitor to zero charge or to any preset output voltage requirement before starting a new integrating process.

5.2 The ideal integrator run, set and hold functions

Figure 5.2 shows how the practical integrator requirements for *run*, *set* and *hold* can be achieved. The manual function switch shown is often replaced by an electronic switching arrangement. In the *run* mode, the circuit produces an

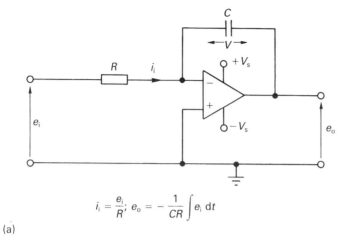

$$i_i = \frac{e_i}{R}; \ e_o = -\frac{1}{CR} \int e_i \, dt$$

(a)

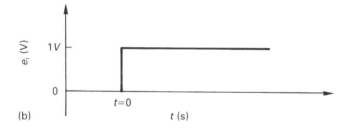

(b)

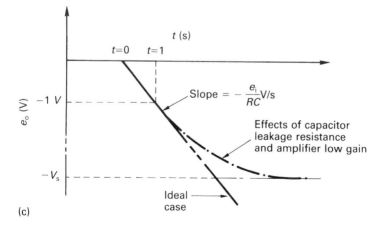

(c)

Figure 5.1 *The basic integrator. (a) The basic integrator circuit. (b) A +1 V step input. (c) The time-related output voltage*

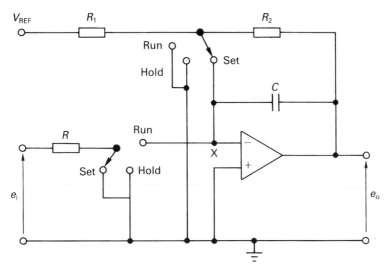

Figure 5.2 *Integrator circuit with run, set and hold facilities. X, virtual earth*

output voltage which is proportional to the time integral of the input signal as illustrated by Figures 5.1(a) and 5.1(b).

In the *set* mode, by using V_{REF}, the output of the integrator can be forced to any desired voltage within the output range of the operational amplifier. The practical limit for the output is usually decided by the d.c. supply voltage, V_s. This is the maximum value to which the output voltage can swing (see Figure 5.1). However, using the *set* function, time must be allowed for the output to change to the new required value; it cannot change immediately. The time taken for the output voltage, which is also that across the capacitor, to change is shown in Figure 5.3. Assuming an initial output voltage level of e_{io} when the *set* voltage is applied, then the capacitor will either start to discharge to the lower set value of e_{os1} or to charge to the higher value of e_{os2} as appropriate.

The upper target voltage is achieved following the curve given by the equation

$$e_o = e_{io} + (e_{os2} - e_{io}) \exp(-t/CR_2)$$

while a similar equation is true for the discharge curve to a lower set output, namely,

$$e_o = e_{os1} + (e_{io} - e_{os}) \exp(-t/CR_2)$$

When switched to the *run* facility, the input voltage is integrated with respect to time and the output voltage is given by the equation:

$$e_o = e_{io} \pm \frac{1}{CR} \int_0^t e_i \, dt$$

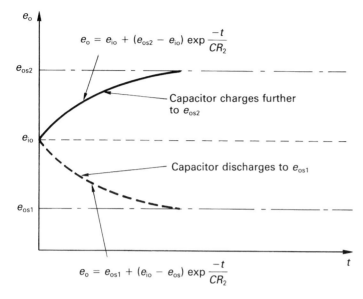

Figure 5.3 *Output volts 'set' action. e$_{io}$, Output volts at the time the output 'set'
voltages e$_{os1}$ or e$_{os2}$ are applied*

When the integrator is switched to the *hold* facility, the integrating action is
stopped and the output voltage of the circuit stays at its then instantaneous
value.

5.3 Practical integrator considerations

When designing practical integrator circuits incorporating operational amplifiers, the inherent errors in these devices must be recognised and allowances
made. The likely problems to be encountered are only summarised below and
for a more detailed treatment the reader should refer to the more specialised
text by Clayton and Newby (1992).

5.3.1 Errors caused by amplifier offset and drift

Even with no applied input signal the integrator amplifier still has to contend
with the input offset voltage, V_{io}, and the input bias current problems, I_b, of
the practical case. These would cause the feedback capacitor to charge continuously. Consequently, the output voltage of a free running integrator would
change continuously until the amplifier output drifted into saturation.

The integrated output voltage drift with time can be eliminated under a
specific set of conditions and at a specific time by using a suitable balance
control to counter the effects of the amplifier input offsets. The problem here is
that having nullified their effects they are still dependent upon a constant

temperature and supply voltage for drift not to reappear. Thus, despite initial balance the integrator output still drifts inexorably to either the positive or negative supply voltage value.

After the initial balancing, the output voltage drift follows a shallow ramp due to temperature changes. The effects of I_b can be reduced by the inclusion of a resistor in the non-inverting terminal connection to earth of the same value as that feeding the capacitor. The input current error is reduced to being caused by the difference between two terminal input bias currents. Another move is to select a large value for the capacitor. This makes the charging current drawn by the capacitor sufficiently large so as to mask the error effect caused by the input bias current. This leaves the input offset voltage as the main cause of drift. The size of the capacitor is limited, however, by the requirement for a particular CR time constant and yet not having too small a value for R which would overload the signal source. Long-term low drift integrating circuits often use high impedance field-effect transistor (FET) input operational amplifiers because these inherently have a small input current requirement.

The integrator error caused by the amplifier input offset voltage and bias current can be estimated from consideration of the equivalent circuit shown in Figure 5.4. The usual analysis is applied assuming that the amplifier has infinite gain and input impedance. This leads immediately to:

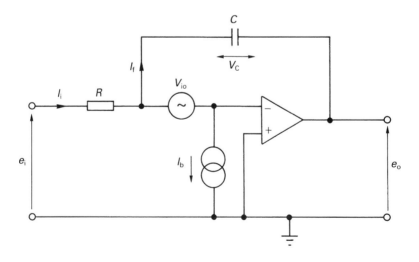

$$e_o = -\frac{1}{CR} \int e_i \, dt \pm \frac{1}{CR} \int V_{io} \, dt \pm V_{io} + \frac{1}{C} \int I_b \, dt$$

Figure 5.4 *Estimating errors caused by input offset and bias current. The first term in the equation is the ideal performance and the last three account for the error caused by the input voltage offset and the bias current*

$$I_f = I_i = I_b$$

where

$$I_i = \frac{I_i \pm V_{io}}{R}$$

but

$$V_c = \pm V_{io} - e_o = \frac{\int I_f \, dt}{C}$$

so

$$-\frac{1}{CR} \int e_i \, dt \pm \frac{1}{CR} \int V_{io} \, dt \pm V_{io} + \frac{1}{C} \int I_b \, dt$$

The first term of the above equation represents the performance of the ideal integrator and the remaining four terms are the error components caused by the input voltage offset and bias current. Use can be made of the equation to estimate the probable errors in a particular application.

5.3.2 Errors caused by finite gain, bandwidth and impedance

In addition to the problems caused by the amplifier input offset voltage and bias current, the practical amplifier has finite values for gain, input impedance and bandwidth. These factors, together with the use of a frequency conscious capacitor having leakage resistance in the operational amplifier feedback loop, limit the accuracy of the integrator response (Clayton and Newby, 1992). One way of looking at the practical integrator circuit is to regard it as an ideal operational amplifier having an infinite gain and infinite input resistance but having a feedback impedance comprising a capacitor in parallel with a resistor. The response of this circuit is discussed in Section 5.4.1.

5.3.3 Slew rate errors

The operational amplifier used in an integrator circuit itself has a finite response time, or output slew rate, for a step input. The amplifier manufacturer's data sheet usually quotes the slew rate for a resistive load. It is the charging time of the amplifier's frequency compensating capacitor which largely determines the slew rate. The integrator goes one step further; it has an additional capacitance in its feedback loop. Since this feedback capacitor is charged by the amplifier output, any limitations in the amplifier's output current capability may result in a slower than expected slew rate. Not only must the amplifier supply the externally connected load, but it has the feedback capacitor to charge at the same time. Suppose the output from the integrator amplifier were limited to i mA. This would be the maximum current which could be diverted to charge the feedback capacitor and presupposes that there are no other load demands on the amplifier. Since a capacitor charges in accordance with the equation $i = -C\,dv/dt$, the capacitor charging rate, which in this case is the amplifier output slew rate, is given by $dv/dt = i/C$. If I were limited to 2 mA and the feedback capacitor were 0.01 μF, the maximum possible slew rate would be 0.2 V/μs.

5.4 Some integrator applications

There are numerous practical applications for the use of an integrating function and the following sections briefly describe only a few.

5.4.1 Simulation of a time lag

Sometimes there is a requirement to design a low cost electronic system which produces conditioned signals to simulate the real performance time lag in a large, expensive and, as yet, possibly unbuilt piece of equipment, for example the time lag between an input change (voltage) and the response (angular velocity) of an electric motor or perhaps the thermal lag in a central heating system.

A circuit which simulates a time lag is shown in Figure 5.5. The principle of operation involved has been mentioned earlier in Section 5.3 where the inadequacies of the practical feedback capacitor with its parallel leakage resistance were discussed. In this particular case the undesirable leakage resistor is purposely augmented by a real one. The operation of the circuit can be considered as that of a simple inverting amplifier having a frequency dependent feedback loop comprising R_f in parallel with C. At low frequencies the capacitor acts as a near open circuit forcing most of the feedback current (ideally the same as the input current) to flow through R_f. The gain of the amplifier at d.c. and low signal frequencies is therefore $-R_f/R$. On the other hand, at high frequencies, the reactance of the capacitance is low, virtually shorting out the effects of R_f and so reducing the amplifier gain to near zero. In theory, a d.c. step input would produce an inverted d.c. step output of magnitude $-e_i R_f/R$. In practice, the maximum output voltage would be limited to the value of supply voltage, $-V_s$, and because the voltage across a capacitor (e_o in this case) cannot be changed immediately, the output voltage would not be able to follow the input step function instantaneously. The output voltage would ramp down at an initial rate of $-e_i/CR$ V/s and gradually flatten out at either $-e_i R_f/R$ or $-V_s$, is whichever the smaller in magnitude.

5.4.2 The summing integrator

Figure 5.6 shows a circuit which makes use of the virtual earth condition of the inverting input terminal to allow several input voltage signals to be added without there being any interaction between their sources. With an ideal amplifier, the currents caused to flow through the three input resistors

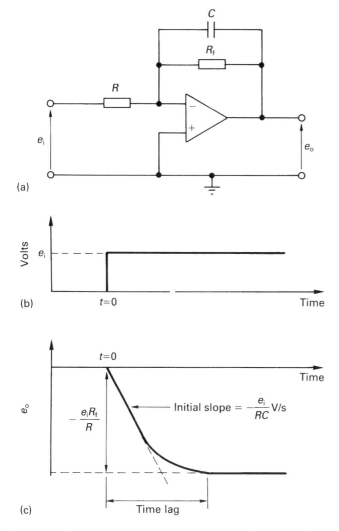

Figure 5.5 *Simulation of time lag showing: (a) the circuit, (b) a step input, (c) the delayed output*

R_1, R_2 and R_3 combine to charge the capacitor C, with the final ideal output being the sum of the separate input voltage time integrals. As before, however, the ideal performance expression must be modified by the inclusion of error terms caused by the unwanted, but unavoidable, effects of input voltage offset and bias current.

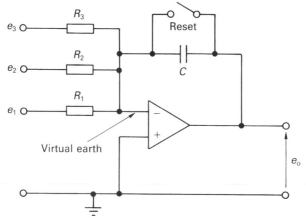

$$e_o = -\left[\frac{1}{CR_1}\int e_1\,dt + \frac{1}{CR_2}\int e_2\,dt + \frac{1}{CR_3}\int e_3\,dt\right] \pm \frac{1}{CR_T}\int V_{io}\,dt \pm V_{io} + \frac{1}{C}\int I_b\,dt$$

Figure 5.6 *The summing integrator. The terms in brackets represent the ideal integrator performance; the remaining terms are the error components caused by the input offset voltage and the bias currents. R_T is the combined input resistance deduced from: $\frac{1}{R_T} = \frac{1}{R_1} + \frac{1}{R_2} + \frac{1}{R_3}$*

5.4.3 The differential integrator

Figure 5.7 shows a single differential input operational amplifier serving as a time integrator of the voltage difference between its two input terminals. The circuit will have an improved rejection of unwanted common mode noise signals if care is taken to match the CR time constants of the circuits associated with the two amplifier inputs.

5.4.4 The current integrator

Should a signal conditioning process require an output voltage proportional to the time integral of a current, the circuit shown in Figure 5.8 can be used. The input current is applied to the basic integrator circuit minus its input resistor. The circuit draws little current from the signal source and therefore produces little loading effect on the circuit being measured.

5.4.5 The current difference integrator

A single operational amplifier can be used to integrate the difference between two currents. No input resistors are fitted. One current is fed directly into the inverting terminal, and the other into the non-inverting terminal. The feedback loop consists of a single capacitor and a second capacitor of equal capacitance is connected between the non-inverting terminal and earth. The problem with

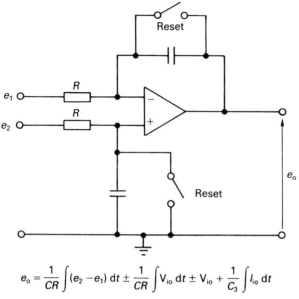

$$e_o = \frac{1}{CR} \int (e_2 - e_1) \, dt \pm \frac{1}{CR} \int V_{io} \, dt \pm V_{io} + \frac{1}{C_3} \int I_{io} \, dt$$

Figure 5.7 *The differential integrator*

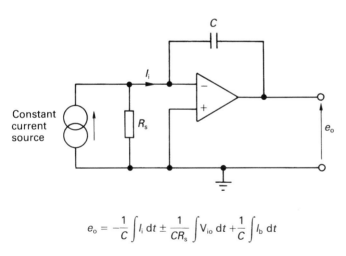

$$e_o = -\frac{1}{C} \int I_i \, dt \pm \frac{1}{CR_s} \int V_{io} \, dt + \frac{1}{C} \int I_b \, dt$$

Figure 5.8 *Current integrator*

this circuit is that any *reset* facility (see the next section) requires that both of the capacitors are simultaneously discharged. For this to function properly two very evenly matched capacitors are needed; so for this and other practical reasons it is usual to use two amplifiers as shown in Figure 5.9. One of the amplifiers acts as a current inverter while the other is a summing integrator.

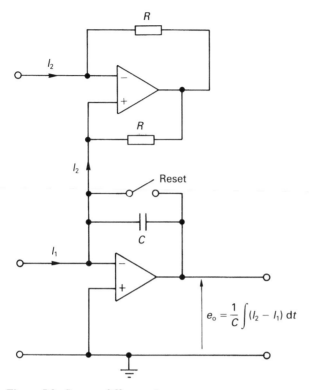

Figure 5.9 *Current difference integrator*

The ideal performance is given by the equation:

$$e_o = \frac{1}{C} \int (I_2 - I_1)\,\mathrm{d}t$$

In practice, this ideal equation requires the addition of the usual error terms to allow for the offset input voltage and the bias currents.

5.4.6 Integrator reset facility

We have seen in earlier sections how an integrator output voltage may be forced to a particular *set* value or *reset* to zero. This latter facility is always required because, unlike with the usual electronic amplifier, the output voltage does not fall to zero on removal of the input signal. Instead, the integrating capacitor remains charged at its instantaneous voltage at the time the input is terminated. The need is to discharge the feedback capacitor and this has been shown as being undertaken by the operation of a manual switch. In practice, it is often more convenient for the capacitor to be shorted out or *reset* by means of a solid state electronic switch arrangement.

A practical low drift integrator (Texas Instruments, 1989) using electronic reset switching is shown in Figure 5.10. The operational amplifier used is a LM 108 and C_1 is the integrating capacitor fitted in the usual feedback position. Metal-oxide semiconductor FETs (MOSFETS) T1 and T2 are triggered into heavy conduction by a negative reset pulse applied to their gates. This discharges (resets) C_1 by effectively connecting both its plates to ground. High input impedance MOSFETS are used in the reset circuitry which is in parallel with C_1. This is in order not to reduce the integrator accuracy by introducing a parallel path for significant capacitor leakage currents. Offset errors are reduced, matching the impedances to ground of the two LM 108 input terminals; R_1 and C_1 are matched by R_4 and C_2. T3 is used to reset the voltage across C_2 to zero.

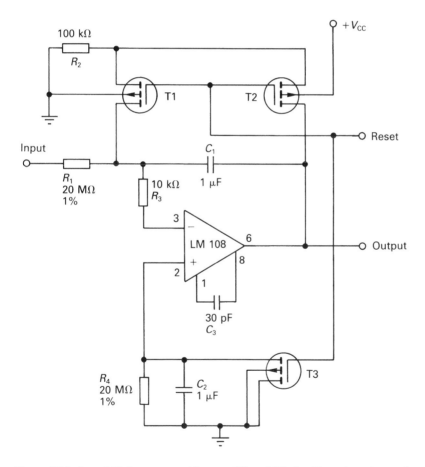

Figure 5.10 *Low drift integrator with reset. T1 and T3 should not have internal gate protection diodes fitted*

5.5 The ideal differentiator

Signal conditioning requirements for differentiation are not so prevalent as those for integration. Differentiators can be used to change the shape of a waveform (see Figure 5.11), to measure the gradient of triangular or sawtooth

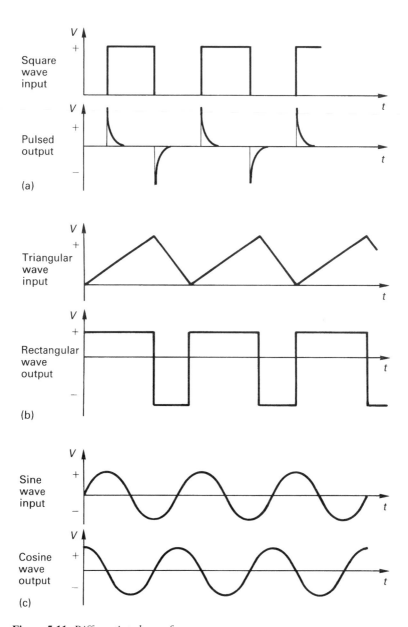

Figure 5.11 *Differentiated waveforms*

waveforms and sometimes to detect a discontinuity in a signal. However, the differentiator circuit has an inherent problem in that, unlike the integrator, it amplifies noise and the performance differences between the ideal and the practical devices are quite pronounced.

A simple differentiator can be produced using an operational amplifier with a capacitor forming the input impedance and a resistor in the feedback loop. This arrangement is shown in Figure 5.12. Assuming an ideal amplifier, all the input current, i_i, can be regarded as also flowing through R_f so that we can say $i_i = i_f$. Using the standard relationship for the current flowing into a capacitor and assuming that the inverting terminal is at a virtual earth potential, we can say that the current $i_i = C\,de_i/dt$. The feedback current $i_f = -e_o/R_f$ so it follows that:

$$C\,de_i/dt = -e_o/R_f$$

whence

$$e_o = -CR_f\,di_i\,dt$$

This shows the output voltage to be the time derivative of the input multiplied by a scaling factor, $-CR_f$.

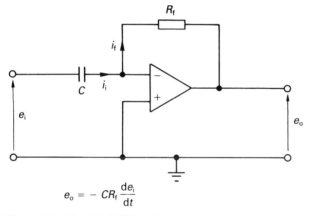

$$e_o = -CR_f\frac{de_i}{dt}$$

Figure 5.12 *The ideal differentiator*

It can be shown that at signal angular frequencies greater than that given by $1/CR_f$, the simple differentiator circuit in Figure 5.12 has a feedback phase lag of about 90° between the amplifier output and the inverting input terminal. This is in addition to the normal 180° inversion in the amplifier itself. Further, for signal frequencies in excess of the limited practical operational amplifier open loop bandwidth, yet another phase lag occurs and the total of the three phase lags can readily come to 360°. This is just the requirement for oscillations to start and the differentiator becomes unstable. Also, at the higher signal frequencies, the differentiator becomes even more liable to instability because of the reduced reactance of the input capacitor, C, causing the differentiator gain to increase.

5.6 Practical differentiator considerations

5.6.1 Gain and frequency limits

In order to produce closed loop stability and some form of high frequency gain limitation, practical differentiator circuits usually use a resistor, R, in series with the input capacitor. This is shown in Figure 5.13. The differentiator gain is now dependent upon the value of R as well as the reactance of C. At the higher signal frequencies of interest, the product of CR can be chosen such that the differentiator gain is limited to R_f/R approximately. This also helps to reduce the undesirable characteristic of the ideal differentiator to amplify extraneous high frequency noise components to the exclusion of the signal itself. A further selected noise reduction can be obtained at the higher frequencies by the fitting of a shunt capacitor, C_f, of appropriate capacitance, as shown in Figure 5.14.

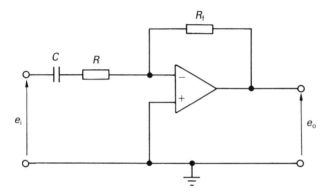

Figure 5.13 *Addition of* R *to improve stability.* R *reduces the bandwidth and high frequency gain of the ideal differentiator*

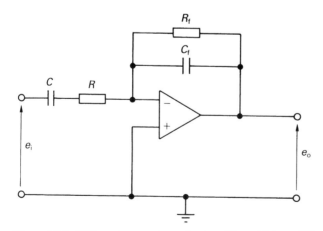

Figure 5.14 *Differentiator noise reduction. The addition of* C_f *reduces the high frequency gain*

C_f reduces the effective impedance of R_f and hence reduces the differentiator gain above the selected frequency.

5.6.2 Offset errors

The usual errors associated with the input offset voltage and bias currents of an operational amplifier exist in the differentiator circuits considered above. Figure 5.15 shows the equivalent circuit used to estimate the extent of these errors. Assuming no current flows into the inverting terminal of the operational amplifier the analysis of the equivalent circuit is quite simple.

$$I_i = C\frac{\mathrm{d}e_i}{\mathrm{d}t}$$

so

$$I_f = I_i = I_b = \pm\frac{V_{io} - e_o}{R}$$

giving

$$e_o = -CR\frac{\mathrm{d}e_i}{\mathrm{d}t} + I_b R \pm V_{io}$$

The first term in the above equation represents the ideal performance and the other two the errors caused by the bias current and the input offset voltage, each of which comprises an initial value and longer term drift. It is usual to balance out the initial values, leaving only the drift problem. The error caused by bias current flow can be reduced by connecting a resistor, which matches the

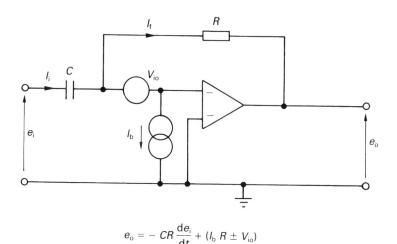

$$e_o = -CR\frac{\mathrm{d}e_i}{\mathrm{d}t} + (I_b R \pm V_{io})$$

Figure 5.15 *Estimation of the differentiator offset error. The term in brackets in the equation is the error voltage*

feedback resistor, between the non-inverting input terminal and earth. This reduces the error current, I_b, to the difference current, I_{io}, between the two amplifier inputs.

5.6.3 Differentiator design considerations

It would appear from the above equation for the differentiator output voltage that C should be chosen to be as large as possible. This would maximise the ideal performance term yet not increase the error terms. Unfortunately, a large value of C requires a small value of R for a given CR time constant. Further, too low a value for R could require the amplifier output to provide a feedback current which, together with its other load current, would be outside its output current rating. Also, the leakage currents associated with large capacitances can add to the existing bias current problems.

A useful approach is to initially select a value for R, in Figure 5.15, in the range 10 to 100 kΩ in order that the amplifier output current limit is not exceeded. The value of C required to produce the maximum output voltage, V_{max}, can then be calculated from the equation:

$$C = \frac{V_{max}}{R\left(\dfrac{de_i}{dt}\right)_{max}}$$

where $\left(\dfrac{de_i}{dt}\right)_{max}$ is decided by the nature of the input signal.

If C is calculated as being greater than 1 μF it may be better to increase the value of R and change the amplifier selection to one drawing a smaller bias current.

5.6.4 The summing differentiator

Figure 5.16 shows how two (or more) input signals can be conditioned by adding them at the amplifier summing point. Each input signal must be provided with its own capacitor which, if required, can be of a chosen value to give the signal a weighting factor. R is included to give the necessary closed loop stability and to reduce high frequency noise.

5.6.5 The differencing differentiator

Figure 5.17 shows a circuit which produces an output voltage proportional to the difference between two input signals. Note how one signal is fed to the inverting terminal and the other to the non-inverting terminal. R is added as before to improve stability and limit noise. In order to balance the amplifier input terminals' d.c. paths to earth, $R_1 = R_f$ is added between the non-inverting terminal and earth.

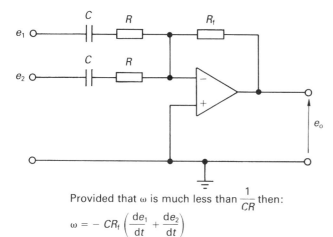

Provided that ω is much less than $\dfrac{1}{CR}$ then:

$$\omega = -CR_f \left(\frac{de_1}{dt} + \frac{de_2}{dt} \right)$$

Figure 5.16 *The summing differentiator*

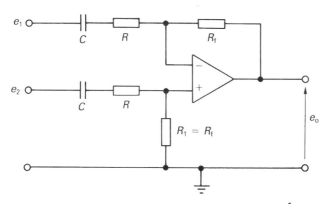

Provided that the signal frequency is much greater than $\dfrac{1}{CR}$ then:

$$e_o = -CR_f \left(\frac{de_2}{dt} - \frac{de_1}{dt} \right)$$

Figure 5.17 *Differencing differentiator*

Exercises

5.1 The integrator circuit shown in Figure 5.1 has $R = 10$ kΩ and $C = 0.1$ µF. Calculate the time taken for the output to become -1 V after the onset of a 10 mV step input.

5.2 Suppose that the operational amplifier in Figure 5.1 has $R = 100$ kΩ, $C = 0.1$ μF and that e_i is a 10 mV step input. Calculate:
(a) the ideal output voltage after 1 ms assuming the input offsets to be zero; and
(b) the total error voltage caused by V_{io} and I_b, if they are 10 μV and 2 nA respectively, and assuming that all the error voltages are in phase. (Hint: see Figure 5.4.)

5.3 An inverting operational amplifier circuit has an input resistor of 50 kΩ and a feedback resistor of 100 kΩ. The non-inverting terminal is earthed. It is required to produce a 1 ms time delay after the application of an input voltage step of 1 V before the output voltage reaches a value of 2 V.
(a) Calculate the value of the capacitor required to be fitted in parallel with the feedback resistor in order to achieve this delay. Assume that the operational amplifier is ideal and that the capacitor charges linearly up to the required output voltage.
(b) To what voltage does the output eventually settle?

5.4 The circuit in Figure 5.13 is supplied by a symmetrical, positive-going, 1 kHz, triangular wave of peak value $+200$ mV and minimum value 0 V. Calculate the maximum and minimum values of the output voltage and draw time-related sketches of the input and output voltage waveforms.

5.5 The circuit shown in Figure 5.12 has $R_f = 10$ kΩ and $C = 1$ μF. If the output voltage, e_o, is not to exceed 5 V with respect to earth, calculate the maximum permissible frequency of a symmetrical triangular input waveform, e_i, having a peak-to-peak value of $+1$ V.

References

Clayton, G. and Newby, B. (1992) *Operational Amplifiers*, 3rd edn, Chap. 6, Butterworth-Heinemann, Oxford.
Texas Instruments (1989) *Linear Circuits Data Book*, Bedford.

6
Switching and waveform generation

This chapter is concerned firstly with the comparator which compares the magnitude of two signals and produces an indication as to which is the larger and secondly with the generation of a range of voltage waveforms that are frequently required by signal processing systems. Both of these techniques call for positive feedback amplifiers either for decreasing switching times or for producing sustained oscillations.

6.1 Ideal comparators

The ideal comparator is a special differential amplifier that is used for detecting and signalling the non-equality of two voltages. We shall see later that the special requirements for the differential amplifier are mainly that it shall have a very high gain, a rapid switching action and a form of hysteresis. The comparator may be used to detect when an increasing voltage, for example a positive or negative ramp voltage, has reached a predetermined level. Also, the comparator output may be used to drive a range of succeeding devices, both analogue and digital.

The simplest form of comparator is shown in Figure 6.1. The amplifier has a very high voltage gain such that a very small differential voltage between the two input terminals will cause the amplifier output to saturate at either its positive or negative supply voltage value. For example, if the amplifier gain were a modest 25 000 and the supply voltage ±15 V, then a difference of 15/25 000 V or 0.6 mV would suffice to make the output swing to the maximum permissible of ±15 V. Clearly, it is a simple matter to design a comparator which will detect voltage differences of only a few microvolts and this is an adequate sensitivity for many processes.

In Figure 6.1, the arrangement is that the voltages to be compared are the reference voltage, E_{ref}, as applied to the amplifier non-inverting terminal, and the signal voltage as applied to the other terminal. With e_i much less than E_{ref} the amplifier output is high at $+V_s$. When e_i is made to exceed E_{ref} by a fraction of a volt, the ideal comparator amplifier will instantaneously change its output from $+V_s$ to $-V_s$. This action is shown in Figure 6.1(b).

Figure 6.2 shows an alternative circuit arrangement where the two voltages to be compared are both fed to one of the amplifier inputs, the other being earthed. This has the advantage of making the comparator immune to common mode voltage (noise) limitations. Because two resistors are used to separate the signal and reference voltage sources, the threshold voltage, V_t, at which the output voltage switches, is given by:

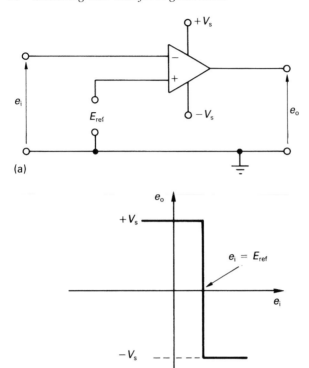

Figure 6.1 *The differential amplifier comparator. (a) Circuit; (b) comparator output switches* $e_i = E_{ref}$

$$e_i = V_t = E_{ref}\frac{R_1}{R_2}$$

The threshold voltage can be adjusted to suit a particular application by choosing the necessary values for R_1 and R_2 and making E_{ref} a convenient voltage of opposite polarity to e_i.

6.2 Practical comparator considerations

If we now take a closer look at the amount by which the input voltage must exceed the threshold voltage for maximum transition of the output to occur, we can see that the output voltage must be made to swing between $+V_s$ and $-V_s$. This assumes that the threshold voltage is zero and the power supply is balanced about zero. For this output voltage swing and for an amplifier having an open loop voltage gain of $-A_{VOL}$, the input signal itself must swing from $-V_s/A_{VOL}$ to $+V_s/A_{VOL}$. The practical amplifier will not have an infinite open loop gain and if the input signal is changing relatively slowly its progress

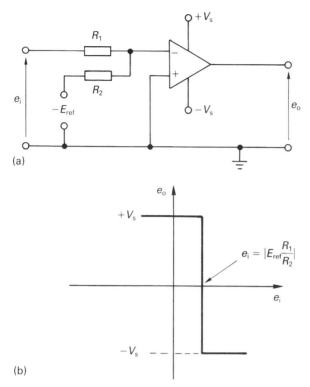

Figure 6.2 *The single ended amplifier comparator. (a) Circuit; (b) comparator output switches when* $e_i = |E_{ref} \frac{R_1}{R_2}|$

through the transition threshold voltage may cause switching uncertainty. This will particularly be the case if the input signal waveform has a shallow slope or has noise superimposed. This situation is illustrated in Figure 6.3. The input waveform shown is triangular but in practice it could be any irregular shape. The amplifier is assumed to have an initial low value for e_i which means that the output is in its high saturated state at $+V_s$. As the input voltage is increased it eventually reaches the threshold voltage at point A and the output starts to fall towards its target of $-V_s$. Unfortunately, in this case, the slow rate of change of the input voltage limits the output switching speed. Further problems could arise if a positive noise voltage spike were to be superimposed on the input waveform just before it reached point A. If the spike were sufficiently large it could cause premature switching by carrying the input waveform past the threshold voltage level. Similarly, a sufficiently large negative noise spike just after point A could cause the comparator to temporarily switch back again.

The above switching uncertainties of the simple comparator are largely overcome by the circuit depicted in Figure 6.4. The secret is the provision of positive feedback to the amplifier through R_2 which, together with R_1, forms a

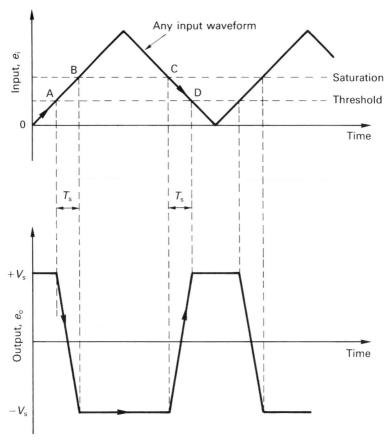

Figure 6.3 *Single ended input comparator switching action.* T_s *is the finite time required to change state. The saturation level is where the input voltage,* e_i, *is sufficient to drive the amplifier into saturated conduction. The threshold voltage* $(= (R_1/R_2)E_{ref})$ *is where the amplifier starts to change state. When* e_i *is less than the threshold voltage, the output voltage,* e_o, *is high*

potential divider between the output voltage, e_o, and E_{ref}. A choice of resistance values for R_1 and R_2 enables a designed value of threshold voltage to be presented to the amplifier non-inverting input terminal. The switching time is much reduced because any initial change in the output voltage is immediately passed back to the non-inverting input which therefore produces a further amplified change in the output and in a 'snowball effect' slams the output in a short time to saturated conduction of the opposite polarity. The target threshold voltage has also been changed in polarity which means that any noise on the input will not be effective until the input has fallen through zero and is about to pass through the new threshold level. A highly desirable form of

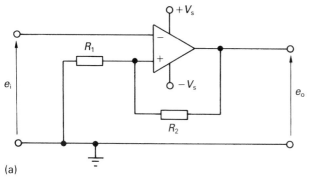

(a)

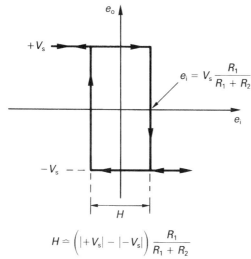

(b)

Figure 6.4 *Comparator with hysteresis. (a) Circuit;* $E_{ref} = 0$. *(b)* H *is the hysteresis voltage*

voltage hysteresis has thus been introduced into the comparator action. This improved switching action is shown in Figure 6.5.

6.2.1 The Schmitt trigger

The operation of this circuit is based on the above hysteresis effect. Figure 6.6 shows a Schmitt trigger circuit which uses discrete components together with its time related input and output voltage waveforms. With no signal present and assuming the transistor V_{sat} to be negligible, transistor T2 is biased into saturated conduction which raises the voltage at point X to approximately 0.455 V; this ensures that T1 is not conducting. With the application of a rising signal voltage, the base of T1 is eventually raised to 1.055 V, that is

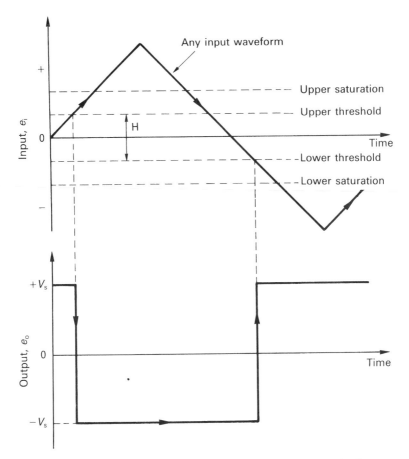

Figure 6.5 *Switching waveforms for comparator with positive feedback. Positive feedback reduces the switching time to approximately zero. H is the hysteresis voltage. Threshold voltages $= \pm V_s[R_1/(R_1 + R_2)]$*

0.6 V above its emitter potential, and T1 starts to conduct. The ensuing initial voltage drop at the collector of T1 is passed through the 10 kΩ resistor to the base of T2 which reduces its current flow accordingly. This constitutes a positive feedback action because the fall in current flowing through T2 is greater than the increase in that flowing through T1 (the different values of collector resistors ensure this) and the net current flow through the 100 Ω joint emitter resistor is reduced. Thus, the voltage at point X falls further, so increasing the current flow through T1 and producing a cumulative effect to cause T1 quickly to switch into saturated conduction and T2 to be turned off. This stable state will exist for as long as the input signal voltage remains above the cut-off level of T1. But the cut-off voltage level of T1 (lower threshold) is now less than its original turn on voltage (upper threshold) because of

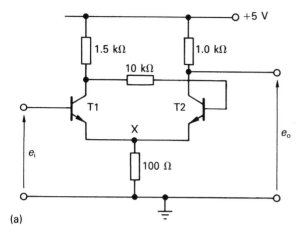

(a)

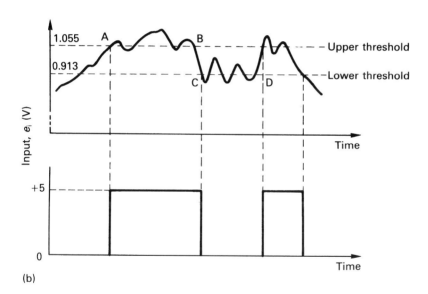

(b)

Figure 6.6 *The Schmitt trigger: (a) the circuit; (b) time related input and output waveforms*

the reduced current passing through the emitter resistor which is now only 5 V/ 1.6 kΩ = 3.13 mA. The emitter potential is now only 0.313 V and the signal must therefore fall to 0.913 V before the conduction switches back to T2. The hysteresis voltage is 142 mV and it is this which prevents any extraneous excursions of the input voltage below the upper threshold (between A and B) or above the lower threshold (between C and D) from causing false switching.

6.2.2 Standard operational amplifiers may not be suitable for all comparator applications

It must be pointed out at this juncture that the standard operational amplifier is unlikely to be suited to all the comparator operations encountered in practical signal conditioning processes. With the application of only a very small differential input voltage, because of its very high gain the standard operational amplifier output voltage readily swings to positive or negative saturation. The output stage is likely to be a push–pull stage and this output voltage swing could typically be ±13 V for a ±15 V supply. If the comparator output is to drive a transistor–transistor logic (TTL) circuit then an output swing between zero and +5 V would be more appropriate and at a transition speed faster than the slew rate of the standard operational amplifier.

Problems such as these have led to the development of special integrated circuit comparators which switch so quickly that the term 'slew rate' is superseded by 'propagation delay versus input voltage overdrive'. The switching rate of the faster comparators is of the order of thousands of volts per microsecond. The term 'overdrive' indicates the excess input voltage above that necessary to cause the comparator to switch into saturated conduction of the opposite polarity. Further, the output circuit of the special purpose comparator chip is very often of the 'open collector' type with an earthed emitter. This allows the collector to be connected to other than the usual (±18 V, say) supply through a resistor (usually called a 'pull up' resistor) to a more suitable voltage supply of, say, +5 V. Figure 6.7 shows such a configuration where a typical open collector output integrated circuit such as an LM311 is used to interface an analogue input to a TTL logic system.

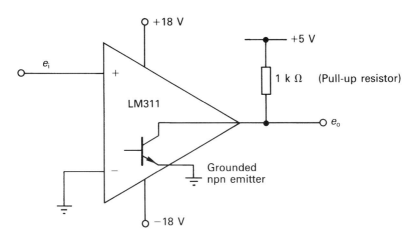

Figure 6.7 *Typical comparator chip with an 'open collector' connected for a TTL logic drive role*

6.2.3 Practical strobed integrated circuit comparator

The LM311 is a typical comparator integrated circuit which has a strobe facility, details of which are shown in Figure 6.8, which is an extract from a manufacturer's data sheet. A block diagram of the device is shown in Figure 6.8(a). The complete chip comprises a single high speed comparator designed to operate over a wide range of power supplies used by operational amplifiers (±15 V) and TTL logic gates ($+5$ V). The output voltages are designed to be used with both TTL and metal-oxide semiconductor (MOS) circuits. The LM311 is also capable of driving lamps or relays and switching voltages up to 50 V at 50 mA. Offset balancing is facilitated as is a strobe input which if taken to a logic low will suppress the comparator output whatever its input. Figures 6.8(b) and 6.8(c) show the output response and the test circuit used for different amounts of input voltage overdrive; the greater the overdrive, the faster the switching.

6.3 Multivibrators

Multivibrators are devices which are much used for the generation of square, rectangular and triangular voltage waveforms. Their operation is usually centred on the charging and discharging of capacitors. There are three common groups of multivibrator: free running or astable (no stable state); monostable (one stable state); and bistable or flip-flop (two stable states). The astable device has an output which continuously switches between a fixed high and a fixed low voltage, at each of which it remains stable for an adjustable short time. This rectangular output is much used for the generation of ramp voltages, after being integrated, or short duration voltage 'pips', after being differentiated. The monostable multivibrator has one permanent stable output voltage from which it is switched or triggered for a short period of time before it automatically relaxes to its original stable state. The bistable device, much used in microelectronic memory devices and registers, has two stable outputs. It will remain in either of these two states until it is triggered into changing. While these devices can be obtained in integrated circuit form, they can be made using ordinary operational amplifiers with external capacitors and positive feedback to stimulate rapid interstate switching.

6.3.1 Astable multivibrators

The circuit for this device is shown in Figure 6.9. The amplifier has a very high gain so that, on switching on, any differential voltage appearing momentarily between X and Y will drive the output into the appropriate positive or negative saturation voltage level. The action of the circuit can be understood by referring to the circuit waveforms shown in Figure 6.10. Assume that at the time of switch on, T_o, the output voltage, e_o, is at its negative saturation level of V_{osat}^-. Then the capacitor, C, charges towards V_{osat}^-, making point X progressively

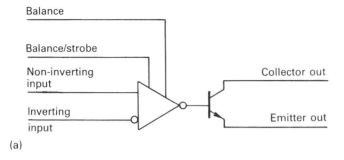

(a)

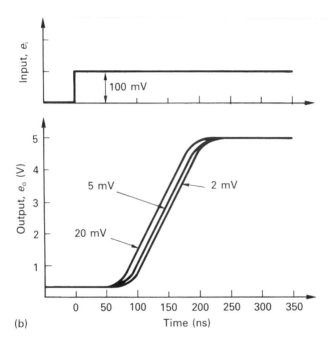

(b)

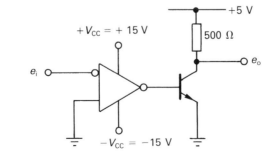

(c)

Figure 6.8 *The LM311 comparator. (a) The functional block diagram. (b) Output response for various input overdrives. (c) The test circuit for (b)*

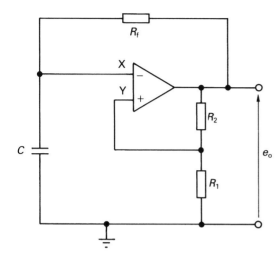

Figure 6.9 *The astable multivibrator*

more negative. While this is happening, point Y on the non-inverting terminal is resting at the reduced value of the output voltage, namely, $\beta V_{\text{osat}}^{-}$, where:

$$\beta = R_1/(R_1 + R_2)$$

Eventually the voltage at X subsides below that at Y and the high gain amplifier together with the positive feedback through R_2 switches the output voltage to the positive saturation level of V_{osat}^{+}. This is now the new target voltage to which the capacitor starts to charge, the actual charging voltage magnitude being the difference between $\beta V_{\text{osat}}^{-}$ and V_{osat}^{+}. The capacitor voltage never reaches its target because before that time it raises the potential at X above that at Y and the output voltage plummets from the positive to the negative saturation level. The capacitor now starts to charge to this lower voltage to repeat the same cycle of events. The output voltage, e_o, therefore rests briefly at V_{osat}^{+} for time t_1 before switching to V_{osat}^{+} for time t_2. The output waveform is thus a square wave because the times t_1 and t_2 are equal, both being dependent upon the capacitor charging time constant which is the product CR_f. The voltage across the capacitor approximates to a triangular wave of peak-to-peak value $2V_{\text{osat}}$.

The frequency of the square wave output is determined by the time the capacitor, C, takes to charge through the resistor, R_f. The standard equation for the time taken for a capacitor C to charge from an initial voltage V_1 through a resistance R_f towards a target voltage of V_2 to reach an intermediate voltage V_3 is given by the relationship:

$$t = CR_f \log_e \frac{V_2 - V_1}{V_2 - V_3}$$

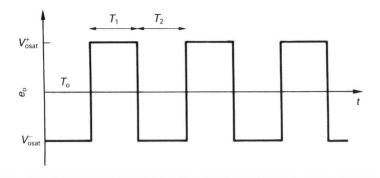

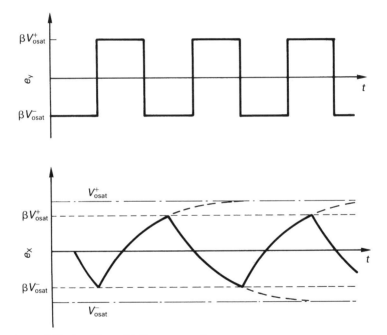

Figure 6.10 *Astable multivibrator waveforms*

Substitution of the appropriate values from Figure 6.10 enables expressions for the times T_1 and T_2 to be deduced as follows:

$$T_1 = CR_f \log_e \frac{V_{osat}^+ - \beta V_{osat}^-}{V_{osat}^+ - \beta V_{osat}^+}$$

$$T_1 = CR_f \log_e \frac{V_{osat}^+ - \beta V_{osat}^-}{V_{osat}^+ - (1 - \beta)}$$

and

$$T_2 = CR_f \log_e \frac{V_{osat}^- - \beta V_{osat}^+}{V_{osat}^- - \beta V_{osat}^-}$$

$$T_2 = CR_f \log_e \frac{V_{osat}^- - \beta V_{osat}^+}{V_{osat}^+ (1 - \beta)}$$

Should the positive and negative amplifier saturation voltages be equal then T_1 will be the same as T_2. We can then say

$$T = T_1 + T_2 = 2CR_f \log_e \frac{1 + \beta}{1 - \beta}$$

which becomes

$$T = 2CR_f \log_e \left(1 + 2\frac{R_1}{R_2}\right)$$

The periods T_1 and T_2 may be varied by adjustment of the capacitor charging time. This can be done by replacing the single feedback resistor by two, each having different ohmic values and each being in series with a directional diode. The circuit arrangement is shown in Figure 6.11 and the associated voltage waveforms are shown in Figure 6.12. When e_o goes positive, C charges through D_1 and R_3 and e_X follows the charging curve AB. During this time diode D_4 is reverse biased and does not conduct. At the end of T_1, decided by the time constant CR_3, e_o switches to a negative value and e_X follows a steeper charging path, BC, because D_4 now conducts and the time constant is reduced to CR_4. R_4 has a smaller resistance than R_3.

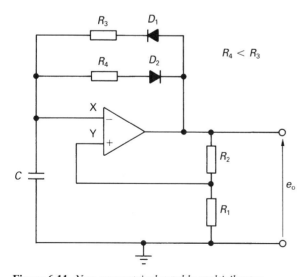

Figure 6.11 *Non-symmetrical astable multivibrator*

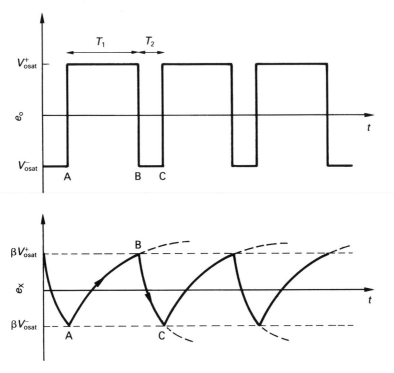

Figure 6.12 *Non-symmetrical astable multivibrator waveforms. A–B has a time constant C_1R_3; B–C has a time constant C_1R_4*

6.3.2 Monostable multivibrators

The directional property of diodes when used with an operational amplifier makes possible the monostable multivibrator circuit shown in Figure 6.13. Ignoring the 0.6 V diode drop, the fitting of D_1 in parallel with the timing capacitor C_1 ensures terminal X cannot be raised above earth potential. Therefore, C_1 cannot be charged positively with respect to earth. The permanently stable state of this circuit is with e_0 at its positive saturation value, making terminal Y also positive but at the reduced value of βe_0. However, terminal X cannot rise to match the Y terminal potential, as would be expected by normal operational amplifier action, because of D_1 acting as an earth clamp. The circuit will remain in this stable state until the Y terminal potential is made negative with respect to X. This latter state can be achieved by applying a sufficiently large negative potential to terminal Y to overcome the existing positive potential of βe_0. Figures 6.13 and 6.14 together show how this can be achieved by the application of a negative square pulse to the differentiating circuit comprising C_2 and R_3. The leading edge of the square pulse produces a large negative-going voltage spike which is applied through D_2 to Y. This switches the circuit output to its negative saturation value, to which the

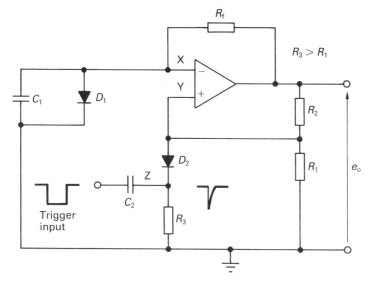

Figure 6.13 *The monostable multivibrator*

capacitor C_1 starts to charge. After an initial small vertical drop the capacitor voltage increases negatively at a rate dependent upon the time constant $C_1 R_f$. But before reaching its target it reaches the same potential as that at Y and switching again occurs. C_1 now charges towards a new target voltage, the positive saturation output voltage, but settles at zero volts because of the earth clamping effect of D_1. The circuit is once again in its permanently stable state where it will remain until triggered.

The time taken by the circuit after being triggered to relax back into its stable state can be estimated using the general expression for the charging of the capacitor C_1 through R_f (see the previous section):

$$T = C_1 R_f \log_e \frac{V^-_{\text{osat}} - 0}{V^-_{\text{osat}} - \beta V^-_{\text{osat}}}$$

so

$$T = C_1 R_f \log_e \frac{1}{1 - \beta}$$

Substituting $\beta = R_1/(R_1 + R_2)$ gives

$$T = C_1 R_f \log_e \left(1 + \frac{R_1}{R_2}\right)$$

Figure 6.15 shows an alternative method of producing a monostable output. The high gain amplifier is used as a comparator. With no trigger signal present, terminal X is pulled down to $-V_{\text{ref}}$ through resistor R_1. Therefore, the output e_o is high at its positive saturated value, the capacitor C charged to e_o and

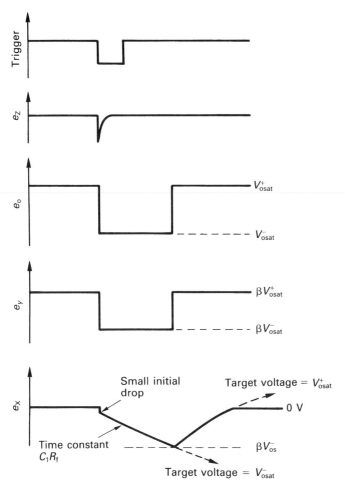

Figure 6.14 *The monostable multivibrator voltage waveforms*

terminal is Y pulled down to earth through R_2. This is the circuit's permanently stable state, in which it remains until a sufficiently large positive trigger pulse is applied to terminal X, making it more positive than terminal Y. This causes the amplifier to switch its output voltage from its positive to its negative saturated voltage. The output fall, V_d, is the difference between these two voltage levels. The voltage fall at the amplifier output is passed straight through C (a capacitor's voltage cannot be changed immediately) to terminal Y. Terminal Y voltage then attempts to recover to earth potential through R_2 on a charging curve determined by the time constant CR_2. Before it reaches its target it passes through $-V_{ref}$, the same potential as terminal X. Immediately this happens, the output voltage is switched back to its high saturated stable state to await the onset of the next trigger pulse. Terminal Y receives a voltage uplift of V_d which

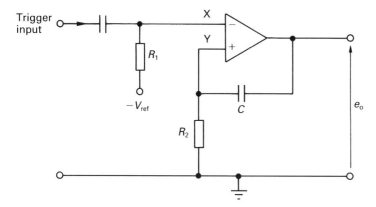

Figure 6.15 *Monostable multivibrator – alternative timing method*

carries it above earth potential only to relax to earth under the CR_2 time constant. Figure 6.16 shows the various terminal waveforms for the complete operation. Using the same approach as previously it can be shown that the period time, T, of the monostable excursion is given by:

$$T = CR_f \log_e \frac{V_{osat}^+ - V_{osat}^-}{V_{ref}}$$

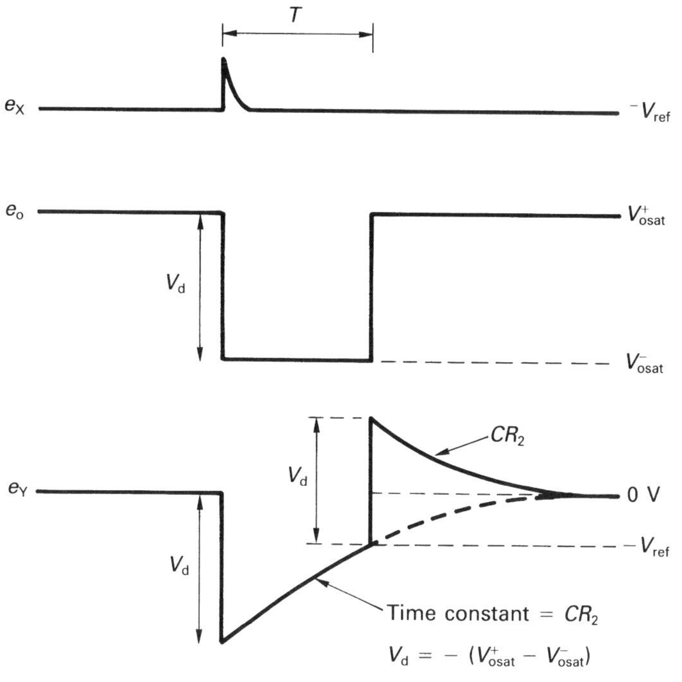

Figure 6.16 *Waveforms for Figure 6.15*

6.3.3 The flip-flop or bistable multivibrator

Figure 6.17 shows a comparator amplifier, with four externally connected components, used to produce the ubiquitous flip-flop or bistable circuit. This is a popular device in the field of digital electronics where it is often used, for example, as a buffer store or as a memory. On switch on, with terminal X connected to earth through resistance R, the output voltage from the amplifier will go high to its positive saturation level. In the absence of a trigger pulse it will remain at this stable state. If now a sufficiently large positive pulse is applied to X then the amplifier output will switch rapidly to its negative saturation voltage level and terminal Y will become a lesser negative voltage given by $\beta = R_1/(R_1 + R_2)e_0$. This is the circuit's alternative stable condition, where it will remain until a negative pulse sufficiently large to exceed the negative potential on Y is applied.

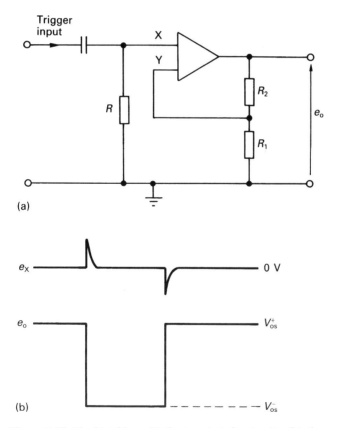

Figure 6.17 *The bistable multivibrator: (a) the circuit; (b) the waveforms*

6.4 Sine wave generators

We have seen in the previous sections how positive feedback is used with multivibrator circuits to speed up the switching of the amplifier output between its positive and negative saturation voltages. Positive feedback is again used, but in a slightly modified way, to produce a continuous switching, or more accurately, a smoothly continuous sinusoidal waveform. The sine wave generator system, which involves self-sustained oscillations, is in effect no more than an amplifier which passes back a fraction of its output voltage to provide its own input requirement for that particular output. This situation would arise, for example, if an amplifier of gain 100 had 1% of its output fed back to its input. Put another way, the loop gain of the system must be unity, that is $\beta A = 1$. If βA is less than unity, oscillations will not occur; if βA is greater than unity the oscillatory output waveform will not be sinusoidal, owing to distortion.

There are many electronic circuit designs which will produce a sinusoidal oscillation. Some of these designs involve the use of tuned circuits comprising inductances and capacitances. Because inductances are expensive and bulky, the circuits which use simple, cheap capacitances and resistors and yet produce satisfactory results are very popular. The 'plug-in' integrated circuit operational amplifier or comparator when used with an external frequency sensitive capacitor–resistor feedback network makes an economical signal generator. The feedback network is designed to produce an overall loop gain of unity at only one frequency. The circuit must also take care that the amplitude of the output does not grow because this will lead to distortion. Similarly, the phase shift through the feedback loop should always be such as to ensure an input back to input loop shift of 360°. The first specific sine wave generator we shall consider uses a Wien bridge as its frequency controller.

6.4.1 The Wien bridge oscillator

Figure 6.18 shows the basic arrangement of the Wien bridge. The capacitors have the same capacitance and the resistances are equal. The circuit forms a potential divider comprising Z_1 and Z_2 across the input voltage, e_i. The output voltage is given by the expression:

$$e_o = e_i \frac{Z_1}{Z_1 + Z_2}$$

If the complex quantities for Z_1 and Z_2 are substituted into the above expression, which is then mathematically manipulated into polar form, it becomes:

$$e_o = \frac{1}{3} e_i \angle 0°$$

This, of course, is for one frequency only given by:

$$f_o = \frac{1}{2\pi C R} \text{ Hz}$$

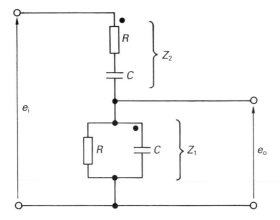

Figure 6.18 *The Wien bridge circuit.* $e_o = \frac{1}{3} e_i$ *and is in phase with* e_i

The circuit in Figure 6.19 shows the Wien bridge together with an operational amplifier acting as a sinusoidal signal generator. Since Z_2 has twice the magnitude of Z_1 then in order to reduce the loop gain to unity the feedback resistors have the relationship $R_2 = 2R_1$. In a practical circuit, for amplitude stability purposes, resistor R_1 is often a thermistor which increases its resistance should the current flowing through it increase. This would be the case should e_o increase. However, the corresponding increase in R_1 would also increase the degree of negative feedback and produce a compensating reduction in the amplifier gain. The signal amplitude would therefore be maintained constant.

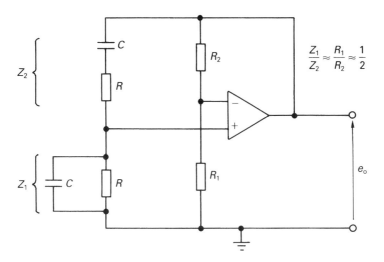

Figure 6.19 *The Wien bridge oscillator*

6.4.2 The sine and cosine wave oscillators

Figure 6.20 shows two amplifiers connected in series and which together produce simultaneous but separate sine and cosine wave outputs. The two outputs are effectively a pair of sine waves separated by a 90° phase difference. The two amplifier stages each form an integrator: one inverts its input; the other has a non-inverting action. The cosine output, e_c, from the inverting stage is fed back to the non-inverting stage input which is also used to provide the sine output e_s. The usual charging current equations for the integrating capacitors, C, associated with two amplifiers can be manipulated to form the following two differential equations:

$$e_c = RC\frac{de_s}{dt}$$

$$e_s = RC\frac{de_c}{dt}$$

Solving these equations for e_s and e_c results in a sinusoidal oscillation at a frequency given by:

$$f = \frac{1}{2\pi CR} \text{ Hz}$$

The Zener diodes associated with amplifier A_2 serve as output amplitude limiters and so provide a form of amplitude stability.

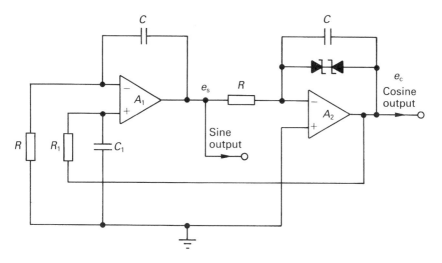

Figure 6.20 *Sine and cosine wave oscillator*

6.5 Function generators

Signal conditioning systems sometimes need the use of other than sinusoidal voltage waveforms. Function generators supply this need. They are available commercially and readily provide such wave shapes as square, rectangular, triangular, sawtooth and the like. However, because there are occasions when a simple waveform generating circuit needs to be built into a signal conditioning system the remainder of this chapter is devoted to a brief description of some of the techniques involved

One of the most popular methods of producing a particular wave shape is to use the readily available integrated circuit operational amplifier as the basic building block. In fact we have seen a form of function generator when we looked at the operation of the astable multivibrator in Section 6.3.1. The main output of this device is a square wave and a triangular wave can be obtained from the timing capacitor. Unfortunately, any attempts to extract a triangular wave from the capacitor is liable to load the circuit such that the frequency is changed or it ceases to function altogether. A more practical approach still is to use the time taken to charge a capacitor for frequency control and to use a comparator action for switching between high and low voltages. The best results are usually obtained by using two operational amplifiers, one for the triangular timing wave generation and the other for the comparator function.

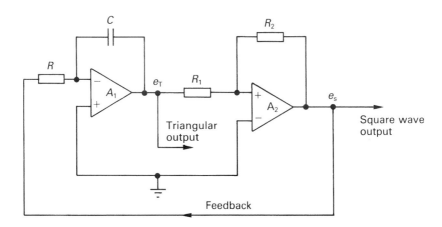

Figure 6.21 *Triangular–square wave generator*

6.5.1 A combined triangular–square wave oscillator

Figure 6.21 shows the typical arrangement for a triangular–square wave generator. Amplifier A_2 is connected as a comparator, the switched square wave output of which is fed back to the input of the integrating amplifier A_1. The output of A_1 is the triangular wave integral of its square wave input. The output of A_2 switches between its positive and negative saturation voltage values:

$$e_s = V^+_{osat} \quad \text{or} \quad e_s = V^-_{osat}$$

On receipt of its square wave input, the inverting integrator output runs down or up, as appropriate, at a rate given by:

$$-\frac{V^+_{osat}}{CR} \quad \text{V/s} \quad \text{or} \quad +\frac{V^-_{osat}}{CR} \quad \text{V/s}$$

The ramping of the integrator output ceases when it reaches the appropriate switching voltage given by:

$$e_T = V^+_{osat} \frac{R_1}{R_2}$$

The periods, T_1 and T_2, for the waveforms (Figure 6.22) are determined by the expressions:

$$T_1 = \frac{(V^+_{osat} - V^-_{osat})\dfrac{R_1}{R_2}}{\dfrac{V^+_{osat}}{CR}}$$

$$T_2 = \frac{(V^+_{osat} - V^-_{osat})\dfrac{R_1}{R_2}}{-\dfrac{V^+_{osat}}{CR}}$$

If we assume ideal operational amplifier action and that the comparator output limits are equally disposed about earth, the two periods T_1 and T_2 are equal and the frequency of the output waveform can be estimated from the following expression:

$$\text{Frequency} = \frac{1}{T_1 + T_2} = \frac{R_2}{4R_1\,CR}$$

6.5.2 Controlled variation of the waveform shape

The geometrical appearance of the waveforms generated by the circuit shown in Figure 6.21 can be changed by modifications to the circuit. The modified circuit which will facilitate these changes in a controlled manner is shown in Figure 6.23. This circuit will allow the adjustment of waveform frequency, its mark-pace ratio or symmetry, the d.c. level of the triangular wave and the triangular wave amplitude.

Varying the value of the timing resistor R controls the frequency of the oscillations. Potentiometer P_1 is used to adjust the voltage V_1 applied to terminal T of the comparator amplifier A_2. This changes the value of both of the comparator switching points by the same amount, V_1/β, where β is the

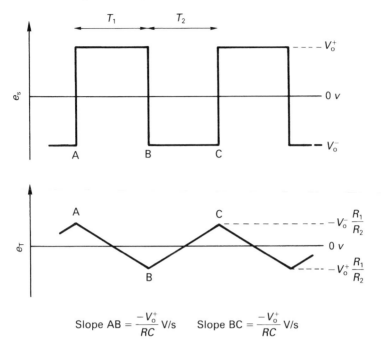

$$\text{Slope AB} = \frac{-V_o^+}{RC} \text{ V/s} \qquad \text{Slope BC} = \frac{-V_o^+}{RC} \text{ V/s}$$

Figure 6.22 *Waveforms for Figure 6.21*

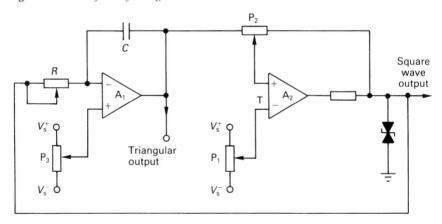

Figure 6.23 *Waveform generator with parameter control. P_1 controls the triangular wave d.c. level; P_2 controls the triangular wave output; P_3 controls the mark-space (symmetry) ratio; R controls the waveform frequency*

positive feedback fraction set by the potentiometer P_2. The effect is to move the triangular wave bodily up or down by adjusting its d.c. level by $V_1\beta$.

The slider position of potentiometer P_2 sets the amount of hysteresis in the comparator A_2. This affects the switching voltages of A_2 and so controls the triangular wave output amplitude. Note, however, that an increase in the triangular wave amplitude will cause a decrease in the output frequency and vice versa.

Adjustment of potentiometer P_3 varies the d.c. offset to the integrating amplifier A_1. This results in an increase in one timing period at the expense of a decrease in the other. This effects control of the waveform symmetry but again causes changes in the output frequency.

6.5.3 Voltage control of waveform frequency

Figure 6.24 shows a circuit arrangement which provides control of the output waveform frequency by means of an externally applied voltage. The circuit is in effect the same as that shown in Figure 6.21 but with the addition of a four-quadrant multiplier device in the feedback link from the comparator to the integrator. The multiplier could typically be the integrated circuit type AD533 which also has the frequency controlling voltage, V_c, as a second input. The output of the multiplier is the product of the square wave input from the comparator and V_c all times the multiplier scaling factor, S. The target charging voltage for the timing capacitor C through resistance R is therefore dependent upon V_c. The expression for the output waveform frequency for the basic circuit shown in Figure 6.21 is thus modified from

$$f = \frac{R_2}{4R_1\,CR}$$

to

$$f = V_c S \frac{R_2}{4R_1\,CR}$$

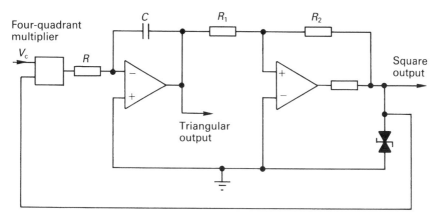

Figure 6.24 *Circuit using a four-quadrant multiplier to effect frequency control of the output waveforms*

Exercises

6.1 For the circuit shown in Figure 6.25(a):
   (a) state the value of e_o when $e_i = 0$ V;
   (b) calculate the value of e_i at which the output will switch, and the value of e_o after switching.

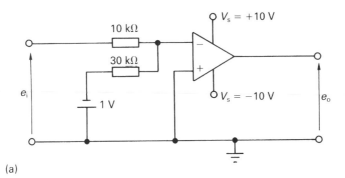

(a)

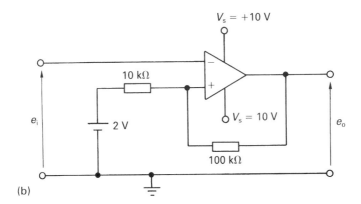

(b)

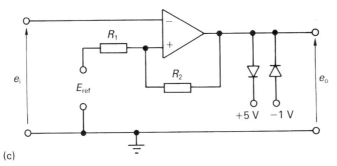

(c)

Figure 6.25 *Comparator circuits: (a) for Exercise 6.1; (b, c) for Exercise 6.2*

6.2 For the circuit shown in Figure 6.25(b), calculate the hysteresis voltage.

6.3 The circuit in Figure 6.25(c) has $E_{ref} = +2.89$ V and $R_2 = 19R_1$. The output voltage is bounded between $+5$ V and -1 V by the diodes D_1 and D_2. Calculate the upper and lower threshold voltages at which the comparator will switch.

6.4 For the circuit shown in Figure 6.9, $R_1 = 20$ kΩ, $R_2 = 100$ kΩ, $R_f = 100$ kΩ and $C = 0.01$ μF.

(a) Calculate the frequency of the output waveform if the output voltage is bounded between $+10$ V and -5 V.

(b) Calculate the percentage change in the frequency if the output voltage were permitted to switch between $+10$ and -10 V.

6.5 Figure 6.26 shows a Schmitt trigger circuit and its input voltage waveform. Sketch the output waveform, e_o, showing the pulse amplitude, duration and recurrence frequency. Assume that the transistors require a V_{BE} of 0.6 V for conduction.

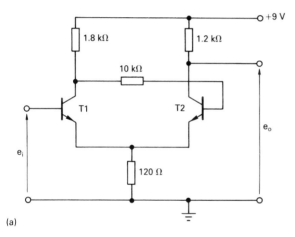

(a)

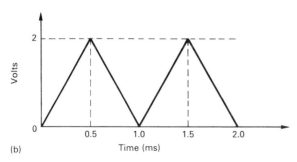

(b)

Figure 6.26 *Circuit details for Exercise 6.5. (a) Schmitt trigger circuit; (b) input waveform, e_i*

6.6 Verify that the output of the Wein bridge shown in Figure 6.18 is one-third the input voltage in magnitude and in phase with it. (Hint: express Z_1 and Z_2 in complex notation form and manipulate these to obtain the modulus and angle of the output voltage.)

7

Analogue processing applications

7.1 Transducer amplifiers

A transducer is the 'front end' or sensing stage of many measuring or instrumentation systems. It is a device for changing one form of energy into another.
This chapter will be concerned only with those transducers which change a
quantity to be measured or processed into electrical energy.

Transducers that change their resistance when stimulated are common. They
include strain gauges, resistance thermometers, thermistors, potentiometric
transducers, light dependent resistors and the like. The usual way of using a
transducer is to place it in a bridge circuit which is adjusted for balance (zero
output voltage) when the transducer resistor is unstimulated. Subsequent
stimulation by mechanical strain, or the application of heat or light, causes a
change in the resistance of the transducer. This unbalances the bridge circuit,
producing an electrical signal proportional to the applied stimulation.

Various forms of resistance bridge circuits are discussed in Chapter 1. Figure
7.1 shows how the relatively small output from a transducer bridge circuit may
be amplified to the level necessary for efficient conditioning. For example, as is
shown in Chapter 10, the conversion accuracy of an analogue-to-digital
converter increases as its input analogue signal magnitude approaches its

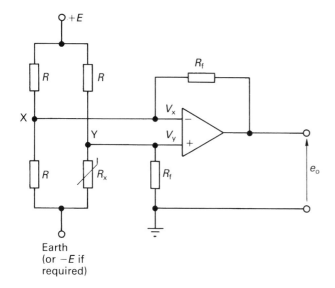

Figure 7.1 *Transducer bridge and amplifier.* R_X *is the transducer element which
changes its resistance from* R *to* $R(1 + a)$ *when stimulated*

designed limit. This may be +2.55 V whereas a typical bridge output is only a few millivolts. In Figure 7.1 the necessary amplification of the bridge output is undertaken by an operational amplifier. The normal operational amplifier feedback action is to make its two input terminals be at the same potential. Thus, the opposing bridge corners at X and Y are also forced to a common potential. The condition for maintaining these equal potentials is produced by the amplifier feedback current which can be analysed as follows.

Equating the currents flowing at junction X and assuming that no current flows into the operational amplifier input terminals:

$$\frac{E - V_X}{R} = \frac{V_X - e_o}{R_f} + \frac{V_X - 0}{R} \tag{7.1}$$

Equating currents at Y and again assuming an ideal operational amplifier:

$$\frac{E - V_Y}{R} = \frac{V_Y - 0}{R(1 + a)} + \frac{V_Y - 0}{R_f} \tag{7.2}$$

Because normal operational amplifier action makes $V_X = V_Y$, if we make both of these equal to V we can rewrite Equations 7.1 and 7.2 to produce the following equation.

$$\frac{V}{R_f} - \frac{e_o}{R_f} + \frac{V}{R} = \frac{V}{R(1 + a)} + \frac{V}{R_f}$$

whence

$$e_o = \left[\frac{V}{R} - \frac{V}{R(1 + a)}\right] R_f \tag{7.3}$$

But we can express V in terms of E and the circuit component values by examining the right limb of the transducer bridge:

$$V = V_Y = E\left[\frac{R(1 + a) \parallel R_f}{R + R(1 + a) \parallel R_f}\right]$$

whence

$$V = \frac{ER_f(1 + a)}{R(1 + a) + R_f + R_f(1 + a)} \tag{7.4}$$

Substituting Equation 7.4 into Equation 7.3 and rearranging we obtain:

$$e_o = \frac{ER_f a}{R} \frac{1}{(1 + a)\left(\dfrac{R + R_f}{R_f}\right) + 1} \tag{7.5}$$

If a is small, Equation 7.5 is virtually linear.

The circuit may be used with an earthed bridge supply. However, it has the disadvantage of having a sensitivity which is dependent upon the bridge impedance levels.

7.2 Resistance measurement

Signal conditioning applications sometimes call for the accurate measurement of an unknown resistance which may be connected to an existing circuit. This *in situ* measurement is possible using an operational amplifier connection as shown in Figure 7.2. The unknown resistance, R_x, neither end of which must be directly earthed, is connected into the feedback circuit of the inverting amplifier configuration. A standard resistor, R_s, of known value is connected at the amplifier input and fed from a reference voltage, E_{ref}. This arrangement forces a constant current through R_x and the value of the amplifier output voltage is directly proportional to the resistance of R_x. Because one end of the unknown resistor is connected to point X, the virtual earth, any other paths to earth from this point have no effect on the reading. With regard to paths to earth from the other end of R_x, then the amplifier output simply provides the extra current required without changing that flowing through R_x.

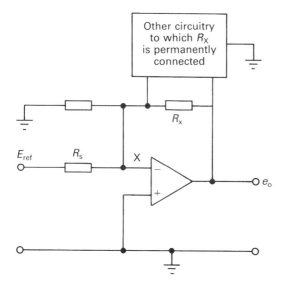

Figure 7.2 *Resistance measurement.* R_X, *unknown resistor for* in situ *measurement;* R_s, *standard resistor; X, virtual earth point;* $e_o = (E_{ref} R_X)/R_s$

7.3 Capacitance measurement

A major problem encountered when attempting to measure small capacitances is the effect of the stray capacitance from the test point to earth. The problem can be alleviated by, once again, making use of the virtual earth property of the inverting operational amplifier. The appropriate circuit is shown in Figure 7.3. The capacitance to be measured is connected to point X and any stray capacitance to earth at this junction is effectively eliminated by having near zero potential across it. The input current from the a.c. signal source, e_i, is passed to

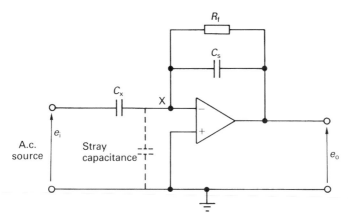

Figure 7.3 *Capacitance measurement. C_X, unknown capacitor; X, virtual earth; C_s, standard capacitor. Voltage gain = $e_o/e_i = (j\omega C_X R_f)/(1 + j\omega C_s R_f)$. Provided the signal frequency is at least $10(1/2\pi C_s R_f)$, then $C_X \approx (e_o/e_i)/C_s$*

the feedback capacitor, C_s, which is a known standard value. For the most accurate results, the chosen value of C_s should be of the same order as C_X and the frequency of the a.c. signal source should be approximately 10 times that given by the expression $1/2\pi C_s R_f$. Under these conditions the amplifier output magnitude is proportional to the ratio of the two capacitor values.

The resistor R_f is fitted in parallel with C_s because unless there is an alternative d.c. path around C_s it is progressively charged by the small amplifier bias current, causing amplifier output drift. This action to prevent drift does introduce a small d.c. output offset because of the small bias current flowing through R_f but this effect is usually tolerable.

7.4 Air speed measurement

Figure 7.4 shows how the passage of cooling air over a sensing element can cause imbalance in an electrical bridge circuit. The complete circuit acts as an anemometer. The circuit is initially set up by adjusting R_2 to produce a suitable temperature in the heated platinum filament sensor. (Platinum is used because it has a large positive temperature coefficient of resistance.) The operational amplifier output voltage, boosted by the emitter follower, automatically assumes the value necessary to ensure voltage parity at X and Y, its two input terminals. Should a cooling airflow now cause the resistance of R_f to fall, the amplifier output voltage will rise in proportion to maintain the temperature of R_f at its original setting. The change in output voltage can be calibrated to indicate airflow. An advantage of this constant temperature method is that accurate measurements of rapid changes in airflow are possible. Because of the thermal delays involved, this would not be the case if a method which depended upon the sensing of a temperature change were used.

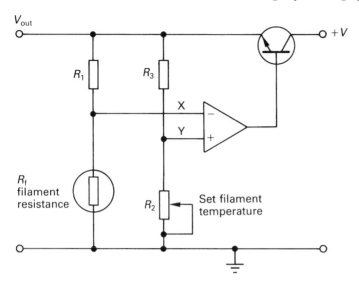

Figure 7.4 *Air speed measurement using constant temperature method. Air flow over hot filament changes R_f from a pre-set value, producing a change in V_{OUT} proportional to air speed over the filament. At bridge balance*
$$R_f = R_1 R_2 / R_3$$

7.5 Capacitance multiplication

Signal conditioning frequently calls for the use of filter circuits or voltage waveform shaping which may require the use of long CR time constants. Because large capacitances are bulky, heavy and expensive, the long time constants are preferably achieved by using a large resistance with a small capacitance. Unfortunately this solution is not successful in all applications. For example, a large resistance may cause biasing, noise or voltage offset problems if used in conjunction with an operational amplifier. If the maximum size of resistance is dictated by other circuit conditions, it is still possible to use a smaller than expected capacitor provided its effective capacitance can be artificially enhanced.

The circuit shown in Figure 7.5 is such a capacitance multiplier. C_1 is the true physical capacitance employed but it is made to appear effectively as C_{eff}, being some C_1 times the gain of amplifier A_2. Thus by varying the value of R_2, and hence the gain of A_2, it is possible to increase the effective value of C_1 at will. The amplifier A_1 is simply a buffer stage to isolate C_{eff} from the loading effects caused by the following inverting amplifier A_2. In practice, the maximum rated output voltage (hence gain) of the amplifier A_2, limits the amount by which C_1 can be multiplied.

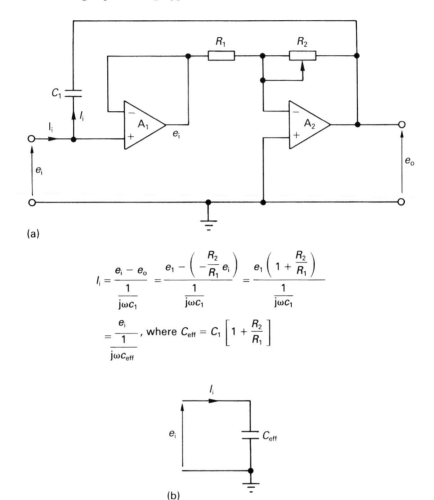

$$I_i = \frac{e_i - e_o}{\dfrac{1}{j\omega c_1}} = \frac{e_1 - \left(-\dfrac{R_2}{R_1} e_i\right)}{\dfrac{1}{j\omega c_1}} = \frac{e_1\left(1 + \dfrac{R_2}{R_1}\right)}{\dfrac{1}{j\omega c_1}}$$

$$= \frac{e_i}{\dfrac{1}{j\omega C_{eff}}} , \text{ where } C_{eff} = C_1\left[1 + \frac{R_2}{R_1}\right]$$

Figure 7.5 *(a) Capacitance multiplier. (b) Equivalent circuit*

7.6 Arithmetic averaging

Arithmetic averaging is a signal conditioning process which produces an output voltage signal proportional to the mean magnitude of a number of input signals. Figure 7.6 shows how the summing point of an inverting operational amplifier is used to interconnect the input signals. The input resistors on each input signal channel are of the same value, R, and the feedback resistor is selected to be R/n, where n is the number of inputs. The amplifier output is the arithmetic mean of the inputs.

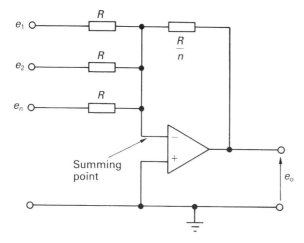

Figure 7.6 *Arithmetic averaging:* $e_o = (1/n)(e_1 + e_2 + \ldots + e_n)$

7.7 Time averaging

Time averaging requires a time or frequency conscious circuit. This is shown in Figure 7.7. To help illustrate the circuit operation, the output voltage waveforms for a square wave input are also shown. The periodic time of the input square wave is very small in comparison with the long time constant of the input low pass filter formed by C and R. After a short time the capacitor charges up to the mean value of the square wave input signal plus and minus a small amount of ripple at double the signal frequency. For correct operation of the circuit it is important that the mean value of the input signal itself does not vary quickly because the long CR time constant will be unable to follow it.

Figure 7.8 shows how the buffer amplifier of the previous circuit can be connected additionally to provide capacitance multiplication. This allows us to use reduced values of capacitance and resistance and yet achieve the long time constant the averaging action requires.

A practical application of the time averaging circuit may be to remove low level, high frequency noise from a much lower frequency wanted signal. The circuit time constant can be selected to follow the variation of the mean of the wanted signal but to be too slow to respond to the high frequency noise which is therefore lost. However, should the mean value of the wanted signal itself suddenly change, for example as it would with a voltage step, then the slow response averaging circuit which removes the high frequency noise would not be capable of following the voltage step. Provided the unwanted noise had a voltage amplitude less than a diode volt drop, say 0.6 V, then the circuit shown in Figure 7.9 would be useful. Diodes D_1 and D_2 connected back to back will short out R_1 if a signal or noise voltage greater than 0.6 V appears at the input. This would be the case for a large voltage step in the input signal, when R_1 would be momentarily removed, so reducing the circuit time constant, allowing the circuit to follow the mean value.

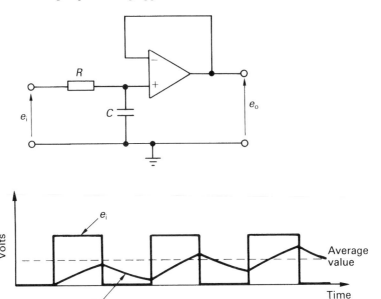

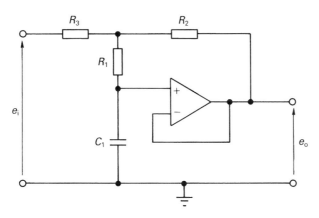

Figure 7.7 *Time averaging:* $e_o = e_i/(1 + j\omega CR)$. *CR, circuit time constant*

Figure 7.8 *Long time constant time averaging using capacitance multiplication:* $e_o = [1/(1 + j\,\omega T)]e_i$; $T = C_1R_1[1 + (R_3/R_2) + (R_3/R_1)]$. *T, time constant*

7.8 Simulation of ideal diode operation

The ideal diode has at all temperatures a zero forward resistance and an infinite reverse resistance. The silicon diode is not ideal; in the forward direction it requires an applied voltage of about 0.6 V before appreciable conduction starts and then the current flow increases non-linearly through a finite resistance; the current flow is very dependent upon the diode junction temperature.

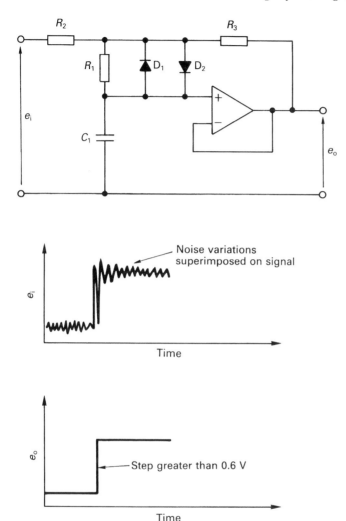

Figure 7.9 *Averaging filter circuit. The time constant,* T, *depends upon the signal amplitude*

Figure 7.10(a) shows a simple practical half wave diode rectifier circuit and the characteristic relationship between the input and output voltages for both the practical and the ideal cases. The difference between the ideal and practical is more pronounced at low input signal levels, where the practical characteristic is markedly non-linear.

However, the circuit in Figure 7.10(b) uses an operational amplifier and acts much like an ideal half wave rectifier. For positive half cycles of input, the amplifier output terminal tends to go negative but diode D_1 is forward biased

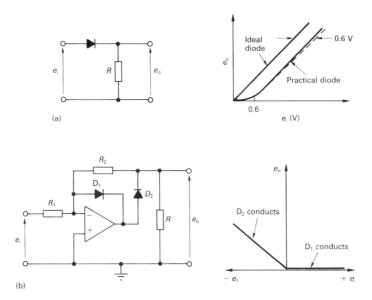

Figure 7.10 *The ideal and practical half wave rectifier characteristics. (b) Operational amplifier circuit acts in the role of an ideal half wave rectifier*

into conduction as soon as the input signal level reaches a level equal to the diode voltage drop divided by A_{OL}. Since the open loop gain of the amplifier may be 10^5 and the diode drop only 0.6 V, the output is virtually clamped to earth. On the other hand, when the input signal is negative, diode D_1 is cut off but D_2 conducts, passing the positive-going output to the load resistor R. The high gain of the operational amplifier also helps to reduce the effects of the non-linear forward resistance of diode D_1. This is because the output voltage magnitude is given by:

$$e_o = e_{in}(R_2 + D_1 \text{ forward resistance})/R_1$$

and since the diode resistance is small compared with R_2 it can be ignored.

The main performance limitation using this type of idealised diode circuit is the finite slew rate of the operational amplifier; this limits the upper frequency response.

7.9 Sample and hold circuits

In digital signal conditioning in particular, it is frequently a requirement for an analogue voltage waveform to be sampled and then held in store awaiting conversion to coded digital pulses. Such a circuit to sample and then hold the sample requires input and output terminals which can be controlled to switch between the 'collect sample' and the 'hold sample' modes. In the sample mode the ideal circuit will follow the input signal it is monitoring and when switched to 'hold' will immediately lock onto, and retain indefinitely, the then present sample level.

In theory, a switch and capacitor is all that is required to undertake this sample and hold function. The circuit arrangement and example waveforms are shown in Figures 7.11(a) and 7.11(b) respectively. When the switch is placed in the sample mode the capacitor is connected across and charges up to the input voltage, e_i. Since the output voltage, e_o, is also connected across the capacitor the output is the same as the sampled input. If the switch is now placed in the hold mode, the capacitor remains charged at the voltage level present at the instant of switching. In practice, the mechanical switch is too slow in operation to allow the necessary rapid sampling of rapidly changing input voltages and the simple capacitor does not retain the sampled voltage for long enough if loaded.

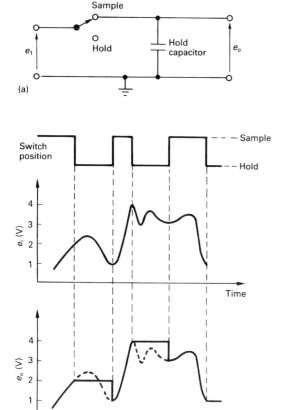

Figure 7.11 *The sample and hold function. (a) Simple ideal sample and hold circuit. (b) Time related waveforms*

Figure 7.12 shows a more practical sample and hold circuit where the switching is undertaken by a field-effect transistor (FET) and the storage capacitor is buffered from the following load by an operational amplifier in a voltage follower connection. This improved sample and hold circuit still does not behave perfectly because it takes a small but finite time for the output to become the same as the sampled input after being switched to the sample mode. This time is called the acquisition time. Similarly, there is a time delay between the circuit being switched from the sample to its settling in the hold mode. This is called the aperture time and should the input signal be changing rapidly and the aperture time relatively long there can be an error between the input signal present and the sample actually held. The accuracy of the circuit in the sample mode is usually expressed in terms of the percentage gain error. The sample and hold circuit is designed to have a unity gain where the input and output voltages are equal.

The most important component in the circuit is the sample storage or hold capacitor. It needs to be sufficiently small rapidly to charge to the level of the sampled voltage and yet it needs to be sufficiently large to retain that level of charge under load in the hold mode. In practice, the final choice of capacitor value becomes a compromise between the requirement for a short acquisition time and a long hold time. Basic sample and hold circuits are readily available commercially in integrated circuit form (LF 198/298/398). All that is required is the connection of an external capacitor of suitable value for the application being considered. The manufacturers of sample and hold integrated circuits publish data sheets giving detailed information regarding the selection of the most suitable hold capacitor for different applications.

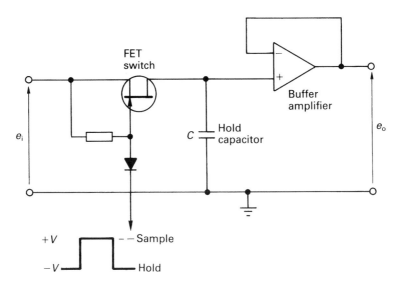

Figure 7.12 *Simple practical sample and hold circuit*

Figure 7.13 shows the basic block diagram of the LF 198 integrated circuit. In order to achieve fast yet accurate sampling rates together with a long hold time, two operational amplifiers are used. Acquisition times can be as low as 4 μs using small hold capacitors while output voltage droop errors can be below 30 μV/s when using large capacitors. The device has a fixed unity gain and an input impedance of 10^{10} Ω independent of the sample or hold mode. The switching logic inputs are at a high differential impedance to permit easy interfacing to any logic family without earth loop problems. A separate offset adjust pin can be used to zero the input offset voltage in either the sample or hold mode. The device can operate over a wide supply voltage range from 5 V to 18 V.

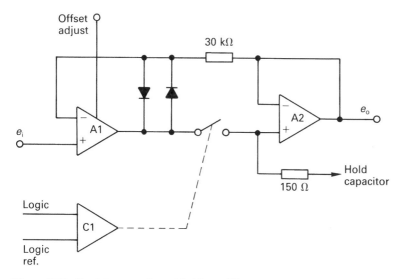

Figure 7.13 *Precision sample and hold amplifier*

7.10 Voltage-to-frequency conversion

The principle involved with one method of converting a d.c. voltage into a frequency is covered in Section 6.4.2. The voltage to be converted is used as the means of controlling the frequency of the output signal from an oscillator comprising an integrator and a comparator. The relevant circuit and its associated waveforms are shown in Figure 7.14. The action of the circuit can be followed by starting with the comparator, A2, output being in its high saturated condition at $+V_s$. The diode D will be heavily reverse biased such that the relatively small positive d.c. input voltage, e_i, will cause the input current all to flow into the integrating capacitor, C. The capacitor charges linearly negatively until its right-hand plate voltage, V_X, reaches $-V_s(R_1/R_3)$, taking V_Y to just below earth. This causes the comparator rapidly to switch over to its negatively saturated voltage output, $-V_s$. Consequently, a heavy current flows through the now forward biased D and the integrator output voltage (right-hand plate

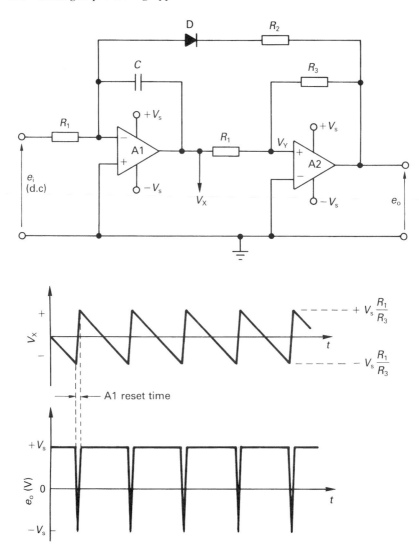

Figure 7.14 *Simple d.c. voltage-to-frequency converter*

of C), V_X, rises positively very rapidly to virtual earth, the potential of the left-hand plate of C. V_x continues its rapid rise through virtual earth and on towards $+V_s$ as the operational amplifier action of A1 maintains the virtual earth potential at its inverting input despite the $-V_s$ being fed back through D and R_2 from the output of A2. The continued rise of V_x towards its target of $+V_s$ is thwarted because when it reaches $+V_s(R_1/R_3)$ the comparator output is switched back to $+V_s$. The cycle now repeats.

7.11 Frequency-to-voltage conversion

Figure 7.15 shows a suitable operational amplifier circuit together with the associated time related voltage waveforms which help in the explanation of its action. The basic operation is one of rectification where the final d.c. output voltage amplitude is proportional to the input signal frequency. The input signal, e_i, is amplified by A1 which has a sufficiently high voltage gain such

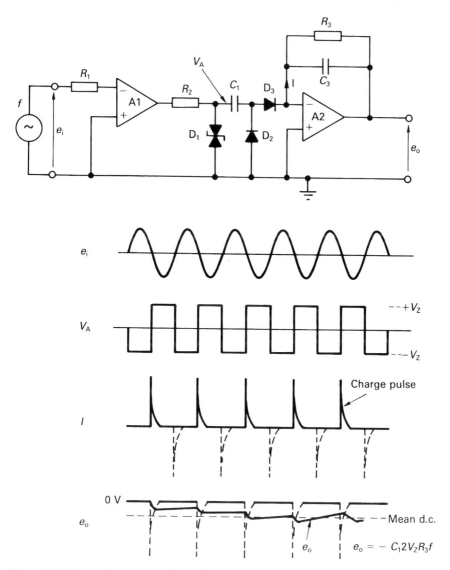

Figure 7.15 *Simple frequency to d.c. voltage converter*

that the following back-to-back Zener diodes, D_1, always conduct, producing a square wave output V_A. The peak-to-peak value of V_A is $2V_Z$ and this is fed into the differentiating section formed by C_1 and the effective following circuit resistance to earth. D_2 cuts off any negative-going pulses, so presenting a train of positive current pulses, at a repetition rate equal to the original input signal frequency, at the inverting input of the integrator, A2. The output of A2 settles to the mean d.c. value of the input pulse train and so is proportional to the original input signal frequency.

Exercises

7.1 A transducer bridge and an operational amplifier are connected as per Figure 7.1. $E = 10$ V, $R_f = 50$ kΩ and $R = 120$ Ω. The transducer has a passive resistance $R_x = 120$ Ω and is stimulated such that its resistance changes by 0.1%. Calculate the amplifier output voltage.

7.2 The circuit shown in Figure 7.3 is used for the measurement of an unknown capacitance, C_x. The standard capacitor, C_s, is 0.1 μF and the feedback resistor, R_f, is 10 kΩ. The a.c. source provides a 1 V input at 400 Hz and the output is measured as 15 V. Calculate the value of C_x.

7.3 A signal conditioning situation requires the use of a special capacitor having a value of 4780 μF. Unfortunately the most suitable capacitor available is only 1000 μF so it is decided to use the circuit shown in Figure 7.5 effectively to increase the value of the capacitor to 4780 μF. Calculate the required gain of the amplifier A2 and suggest suitable values for R_1 and R_2.

7.4 The circuit shown in Figure 7.8 has $R_1 = 100$ kΩ, $R_2 = 1$ kΩ, $R_3 = 10$ kΩ and $C_1 = 0.01$ μF. If the input voltage is a positive-going square wave of amplitude 1 V and frequency 10 Hz, calculate:
   (a) the effective time constant of the circuit; and
   (b) the average value of the settled output voltage.

8

Noise

8.1 Introduction

In virtually every electrical measurement or signal conditioning application, there are unwanted or stray signals which are intermingled with the wanted signal. The unwanted signals, from whatever source, are termed noise. If the wanted signals are very weak they can be completely lost by submersion below a more powerful noise level. This is depicted in Figure 8.1.

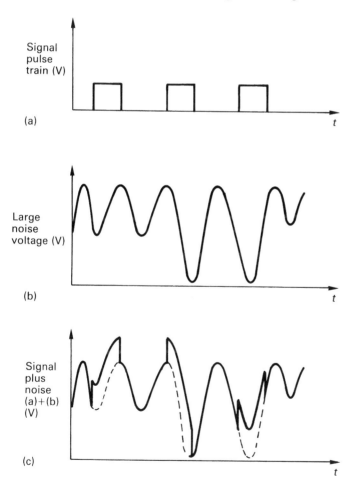

Figure 8.1 *The undistorted signal at (a) when added to the noise voltage at (b) results in the unreadable waveform at (c)*

In general the sources of the unwanted noise power can be classified as internal or external noise. Internal noise is that unwanted random signal power which appears at the output of an electronic device despite there being no wanted signal input; it is taken as being generated within the device itself. On the other hand, external noise power is regarded as the combined unwanted electrical signal powers which are picked up by the device from whatever external source. We shall examine both of these noise sources in more detail later but first we must understand the meaning of the term signal-to-noise ratio.

8.2 Signal-to-noise ratio (SNR)

We have seen above how a weak signal can be lost in a more powerful noise background. The SNR is simply a method of comparing the relative magnitudes of the two at the frequency of interest. The SNR is defined as S/N where S is the wanted signal power present and N is the unwanted noise power present.

It is customary to present the SNR in decibels, so we can say that

$$\text{SNR} = 10 \log_{10}(S/N)\,(\text{dB})$$

For example, suppose the wanted signal power being fed into an amplifier were 3 mW and that this were accompanied by an unwanted noise power input level of 30 µW. The SNR would be 20 dB. While this SNR figure would be acceptable for a telephone link, a high quality music centre would require figures in excess of 60 dB.

A practical point to remember is that the SNR at the input of an amplifier is not improved by the subsequent gain factor of the amplifier. The noise power is amplified in just the same way as is the wanted signal. In fact the SNR will be worse at the amplifier output by the addition of the inevitable internally generated noise.

8.3 Sources of external noise

At the output of an audio amplifier, one form of external noise is heard as a hissing sound commonly known as 'mush'. Unwanted signals are also classified as external noise and these may be heard as anything from a high pitched whistle to a low hum. The latter is particularly characteristic of what is called 'mains pick-up', being an extraneous input from the 50 Hz a.c. mains power supply.

8.3.1 Mains interference

Figure 8.2(a) shows a block diagram of an audio amplifier driving a load. The stray capacitances, C_s, which exist between the mains power supply lines and the signal carrying wires couple (or pass) a small amount of the 50 Hz power

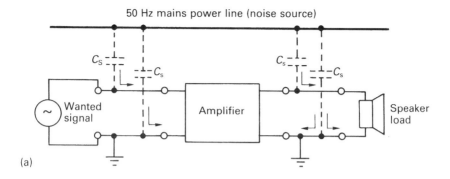

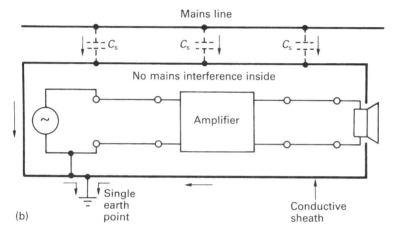

Figure 8.2 *(a) Mains interference coupled into an audio system through stray capacitances, C_s. The arrows show the noise interference paths. (b) A conductive sheath around the sensitive equipment can form an interference shield*

frequency into the audio system. Not only is the fundamental power frequency of 50 Hz injected, but so are its harmonics of 100 Hz, 150 Hz and so on. The low growling or higher pitched humming heard at the system output is usually attributable to this source of noise. The effects of power line interference can be reduced by carefully designing the amplifier layout so the power lines are located as far away from the signal lines as is possible. This not only reduces the coupling effect of the stray capacitances; it also reduces the magnetic coupling which also exists between any pair of current carrying conductors. A more drastic and expensive method of reducing mains interference is to physically shield the whole amplifier and its associated audio system in a conductive metal casing which is earthed at a single common point. Any stray reactive coupling between the power and audio lines is broken and the noise picked up is shorted straight to earth. Figure 8.3(b) shows the noise shielding arrangement.

8.3.2 Other man-made noise sources

Besides the noise interference picked up from nearby mains power lines, there can be much nuisance interference from the radiation caused by electrical machinery. Typical examples of this type of noise are the sparking commutator brushes in many d.c. and universal motors, fluorescent lights, switch gear and any other source which employs the rapid breaking or changing of currents. If the sensitive signal conditioning equipment is concerned with the acquisition of data over audio lines or at radiofrequencies then there are additional sources of noise with which to contend. These are described below.

8.3.2.1 Adjacent channel interference
This is particularly applicable to radio links where data acquisition channels are separated by being designed to operate on different carrier frequencies. Nevertheless, unless there is a considerable geographical spacing, it is possible for the power of the adjacent channel's carrier to be sufficient to allow it to produce interference whistles or even to swamp the wanted channel to the point of exclusion. The answer to this problem is the geographical spacing of equipment operating on adjacent channels plus the use of good frequency selective input circuitry on each channel.

8.3.2.2 Cross-talk
This term originates from the problems that occur when two or more telephone lines or data-carrying cables lie together. Even though the cable conductors are insulated one from the other, the alternating magnetic and electric fields established around the cables by the varying currents they are carrying cause the signal in one cable to be induced into the adjacent ones. The problem is best overcome by using screened cables or by not routing the offending cables close together.

8.3.2.3 Intermodulation noise
It is shown in Chapter 10 that if two signals, one of frequency f_1 and the other f_2, are mixed together in a device which has a non-linear input to output characteristic then the product comprises four signals of frequency $f_1, f_2, f_1 - f_2$ and $f_1 + f_2$. In a radio circuit this is an intended signal conditioning process but in data acquisition and subsequent processing it may happen unintentionally to produce intermodulation noise. The situation can be even more complex if the mixing of the harmonics of the prime signal frequencies is considered. The output may then contain signals of significant amplitude at $f_1 \pm f_2, 2f_1 \pm f_2, 2f_1, 2f_2, 2f_2 \pm f_1, 3f_1 \pm f_2$, and so on. This effect of producing many new but unwanted signal frequencies, some of which will lie in the band-width of the signals of interest, can cause a serious degradation of the SNR.

8.3.2.4 Quantisation noise

This is the result of attempting to reconstitute an analogue waveform from a train of variable height digital pulses. This topic is dealt with more fully in Chapter 11.

8.3.3 Natural noise

These forms of noise, described below, are mainly limited to circuits working in the radio communications frequency bands.

8.3.3.1 Atmospheric or static noise

Radio aerials are particularly susceptible to this form of noise which basically is generated by electric storms in the atmosphere. The electromagnetic waves produced by the lightning discharges propagated powerful noise signals over a wide area. Very often, these unwanted noise signals can temporarily completely obliterate the wanted data signal. The unwanted radiation is experienced up to about 25 MHz

8.3.3.2 Noise from outer space

This is also known as galactic or cosmic noise. It emanates from the stars and includes the interference caused by our own 'sun-spot' activity. It is effective at frequencies beyond 1 GHz.

8.4 Suppression of externally generated noise

Various techniques are adopted in an attempt to remove the nuisance of external interference. The degree of success depends upon the frequency of the interference and its power level and upon the sensitivity of the affected equipment.

8.4.1 Shielding

Figure 8.2(b) shows the basic principle involved with this technique of reducing the amount of noise picked up through stray capacitances and inductances.

Magnetic field interference at d.c. and at low a.c. frequencies · is best overcome using a shield made from a metal having a high magnetic permeability. By virtue of its very low magnetic reluctance, the hi-mu shield forms a better magnetic path for the incoming magnetic interference flux than does the free space within the shield. The sensitive components inside the shield thus remain unaffected.

Electromagnetic wave interference at radiofrequencies (which includes that produced by 'sparking' equipment) is best eliminated using a metal shield surround having a very low electrical resistance. Aluminium, brass and copper are good examples of suitable materials. The principle involved at these higher frequencies is that the magnetic flux component of the interference strikes and therefore 'cuts' the low resistance shield. This constitutes a rapid flux change (at radiofrequency) which, in accordance with Faraday's law, produces an induced eddy current in the surface of the metal shield. The flux produced by the eddy current, by Lens's law, is in opposition to the original flux causing it. If the electrical resistance of the metal shield is very small and the frequency of the interference is very high, then the eddy current flux produced virtually cancels the interference flux and the noise signal is reflected away from the shield. This is another manifestation of the 'skin effect' in that it is the induced eddy currents flowing in the surface or 'skin' of the metal shield which prevent its penetration by the interference noise flux.

Other shielding techniques include the use of braided cables and co-axial cables wherever there is the possibility of unprotected wires being coupled into unwanted radiation. Also, while it is not so effective against magnetic interference, a mesh wire shield can be successful in subduing electrostatic and capacitively coupled noise. The wire mesh screen is not so heavy or expensive as the sheet metal type, and is often used to make a 'local' screen around a particularly sensitive component.

8.4.2 Circuit layout

While it may appear to be stating the obvious, great care should be taken when designing the layout of a piece of electronic equipment to either eliminate or reduce interference pick-up. Mains frequency supply lines should be spaced away from signal input lines as should obvious sources of magnetic radiation such as transformers. Power lines and signal lines should not share the same trunking nor should they share the same access holes through amplifier chasses.

8.4.3 Elimination of earth loops

Figure 8.3 shows the situation which can arise if a signal conditioning system has more than a single earth bond. The case shown is where both the signal source and the amplifier input have been separately earthed. The points A and B may be well spaced, allowing the line AB to act as an unwanted aerial to pick up any noise through inductive or capacitive coupling. Further, the points A and B may well be at slightly different 'earth' potentials because of different earth bonding resistances at A and B. These effects are sufficient to produce unwanted interference noise signals in series with the input signal.

8.5 Sources of internally generated noise

However carefully the suppression of externally generated noise pick-up has been, the ultimate criterion which limits the minimum level of signal which can be detected and conditioned is the internally generated noise. Whether active or passive, all electronic components generate unwanted electronic noise.

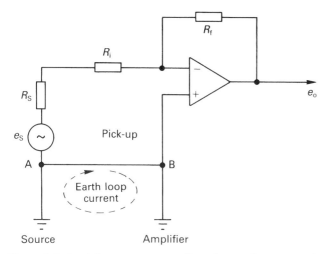

Figure 8.3 *Earth loop currents can flow if more than one earth point exists. Because of stray noise and pick-up, $e_A \neq e_B$*

8.5.1 Thermal noise

This is also known as Johnson noise or white noise. It is caused by the random movement of electron current carriers in both conductors and semiconductors. The amount of electron (and hole) motion is dependent upon the temperature of the component. The higher the temperature, the greater the electron agitation. If the component is considered as being connected into an external circuit, the random drifting of the thermally agitated current carriers will be first in one direction and then in the other, so causing momentary voltage differences to be set up between the component terminals. The random voltage fluctuations so caused will average out to zero over a long period. But their r.m.s. value over the same long period will constitute the generation of a small noise voltage.

Johnson studied this phenomenon and found that the r.m.s. value of the thermal noise voltages, E_n, over a given bandwidth, B Hz, at an absolute temperature, T kelvin, for a given material of electrical resistance $R\ \Omega$ is given by the equation:

$$E_n = \sqrt{(4KTBR)}\ (\text{V})$$

where k is Boltzmann's constant, being 1.38×10^{-23} J/K.

Since (voltage)^2 is proportional to power, it can be deduced from Johnson's equation that thermal noise power is directly proportional to the bandwidth being considered, so an amplifier having its bandwidth doubled will experience a doubling of its noise output. Looked at another way, this means that thermal noise is the same whatever the frequency. Thermal noise is evenly distributed over the frequency spectrum and this is the reason for its being called 'white noise'. The colour white is taken as being composed of all colours of all frequencies.

8.5.2 Shot noise

Shot noise is the effect of the random variations in the flow of electrons at the emitter–base and the base–collector junctions in a transistor. Current is commonly taken as being a smooth fluid-like flow of electrons, but this is not the case. The r.m.s. value of the shot noise current is given by the equation:

$$i_n = 2eIB \, (\text{A})$$

where $e = 1.60 \times 10^{-19}$ in coulombs, I is the emitter d.c. current in amps, and B is the bandwidth under consideration in hertz.

For a given bandwidth, there is a linear relationship between the d.c. current flowing through the transistor and the noise current produced; the more current the transistor is asked to handle, the noisier it becomes. Once again, the noise power is uniform across the frequency spectrum and therefore is classified as white noise.

8.5.3 Flicker noise or $1/f$ noise or pink noise

This arises because of the random rate at which holes and electrons are generated and then recombine in semiconductor devices. Flicker noise tends to be restricted to the lower audio frequencies below 10 Hz. Its magnitude is inversely proportional to frequency; hence the term '$1/f$ noise'. Also, the red colours are in the lower part of the colour frequency spectrum; hence the term 'pink noise'.

8.5.4 Partition noise

This arises in transistors because of the random variation in the way the emitter current divides between the base and collector terminals.

8.6 Worked examples on internal noise calculations

8.6.1 Problem 1

Figure 8.4(a) depicts an amplifier which has a bandwidth, B, of 15 kHz and an input resistance of 1 kΩ. The input signal is provided by a source, e_s, of 1 mV and internal resistance, R_i, also of 1 kΩ. Assume the temperature of the system to be 20°C (293 K) and take Boltzmann's constant, k, as 1.38×10^{-23} J/K. Calculate: (a) the r.m.s thermal noise generated in the source resistance; and (b) the input noise power and input SNR caused by (a).

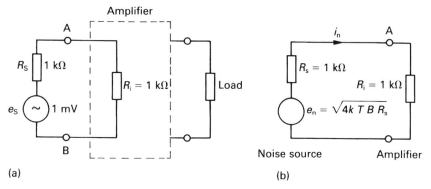

Figure 8.4 *Circuits for the problem worked in Section 8.6.1*

Calculations

(a) The r.m.s. thermal noise generated in a 1 kΩ source resistance is given by:

$$e_n = \sqrt{(4kTBR_s)}$$
$$= \sqrt{(4 \times 1.38 \times 10^{-23} \times 293 \times 15 \times 10^3 \times 10^3)}$$
$$= 0.49\,\mu V$$

(b) Figure 8.4(b) shows how the 0.49 μV noise generated in the resistance R_s can be represented in the system as a noise alternator in series with R_s.

Therefore, the r.m.s. thermal noise current is given by $i_n = e_n/(R_s + R_i)$, thus making the noise power developed in the input resistance

$$N_i = i_n^2 R_i = e_n^2 R_i/(R_s + R_i)^2$$

Because $R_i = R_s = R$, we can rewrite this as

$$e_n^2 R/(2R)^2 = e_n^2/4R$$
$$= (0.49 \times 10^{-6})^2/(4 \times 10^3)$$
$$= 6 \times 10^{-17}\,W$$

Similarly, the input signal power appearing in R_i is

$$S_i = e_i^2/4R$$
$$= (10^{-3})^2/(4 \times 10^3)$$
$$= 2.5 \times 10^{-4}$$

Thus, the input SNR, S_i/N_i, can be written as

$$(2.5 \times 10^{-4})/(6 \times 10^{-17}) = 4.17 \times 10^{12}$$

or, expressed in dB, 10 log(4.17 ×10^{12}) = 126 dB.

8.6.2 Problem 2

An amplifier has a power gain of 10 dB, a bandwidth of 6 MHz and an input resistance of 1 kΩ. It amplifies a signal having a bandwidth of 6 MHz. The signal source has an internal resistance of 1 kΩ and a temperature of 20°C. The total noise generated in the amplifier is equivalent to 0.1 pW referred to the amplifier input. If the signal input power is 1 μW, calculate: (a) the SNR resulting from the noisy source; and (b) the total noise input power, the signal output power and the output SNR.

Calculations
(a) Because the source and amplifier input resistances are both 1 kΩ, the noise at the amplifier input caused by the thermal noise of the source resistance (see the solution in Section 8.6.1), N_i, is:

$$N_i = (e_n)^2/4R = 4kTBR/4R = kTB$$
$$= 1.38 \times 10^{-23} \times 293 \times 6 \times 10^6$$
$$= 2.43 \times 10^{-14}\,\text{W}$$

The signal input power $S_i = 1$ μW and so the SNR is:

$$S_i/N_i = 10^{-6}/(2.43 \times 10^{-14}) \text{ or } 10 \log\,S_i/N_i = 76\,\text{dB}$$

(b) The total noise power output, N_o, is given by:

N_o = power gain × (Source thermal noise + Amplifier internally generated noise

referred back to the input)

$$= 10\,(\text{from 10 dB}) \times (2.43 \times 10^{-14} + 0.1 \times 10^{-12})$$
$$= 1.243 \times 10^{-12}\,\text{W}$$

The signal output power, S_o, is given by:

$$S_o = \text{Power gain} \times S_i = 10 \times 1\,\text{μW} = 10\,\text{μW}$$

This gives a SNR of:

$$S_o/N_o = (10 \times 10^{-6})/(1.243 \times 10^{-12})$$
$$= 8 \times 10^6 \text{ or } 69\,\text{dB}$$

8.7 Maximum available signal power and noise power

The maximum power which can be transferred from a source, of given internal resistance, to a load occurs when the product of (Circuit current)^2× Load resistance is a maximum. Examination of Figure 8.5(a) shows the circuit conditions.

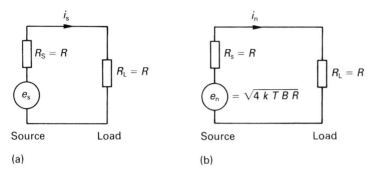

Figure 8.5 *Circuits for signal power and noise power considerations. (a) Maximum power is transferred between source and load when their impedances are the same: $R_s = R_L = R$. Maximum signal power is $S_{Ai} = e_s^2/4R_s = e_s^2/4R$. (b) Thermal noise generated by the source in R_s is $e_n = \sqrt{4KTBR_s}$. Maximum noise power is $N_{Ai} = e_n^2/4R = KTB$ when $R_s = R_L = R$*

The circuit current, i_s, is given by $e_s/(R_S + R_L)$. Therefore the power dissipated in the load is:

$$\text{Power} = (e_s)^2 R_L/(R_S + R_L)^2$$

$$\text{Power} = \frac{e_s^2}{\dfrac{(R_S + R_L)^2}{R_L}}$$

The maximum power dissipation occurs when the above expression is a maximum and this is when the denominator is a minimum. Differentiating the denominator with respect to R_L and equating it to zero shows that the maximum power dissipation occurs when $R_S = R_L$. If we make both of these resistances equal to R we can say that the maximum available power dissipation is:

$$S_A = (e_s)^2/(2R)^2 = (e_s)^2/4R$$

Similarly, with reference to Figure 8.5(b), we can say that the maximum thermal noise transferred from the source resistance to the load, at a temperature of T kelvin, is:

$$N_A = \frac{e_n^2}{4R} = \frac{4kTB}{4R} = kTB$$

8.8 Maximum available power gain

The maximum available power gain, G_A, occurs when the source and amplifier input resistances are the same while the amplifier output and load resistances also match. See Figure 8.6.

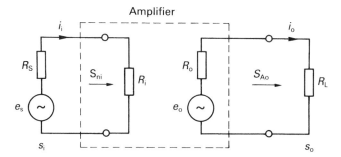

Figure 8.6 *Maximum power gain occurs when* $R_s = R_i$ *and* $R_o = R_L$, *where* R_i *and* R_o *are the amplifier input and output resistances, respectively*

$$G_A = S_{Ao}/S_{Ai}$$

where S_{Ao} is $(e_o)^2/4R_o$, the maximum available output power, and S_{Ai} is $(e_s)^2/4R_s$, the maximum available input power.

8.9 Noise temperatures

In order that the noise generated by a circuit can be represented in terms of a temperature, it has been agreed that such noise calculations would be undertaken assuming a standard ambient temperature of $T_0 = 290$ K. Therefore, assuming that $k = 138 \times 10^{-23}$, this means that $kT_0 = 41 \times 10^{-21}$ Ws.

The effective noise temperature, T_{eff}, of a source is defined as the product $(1/kB) \times$ (Available noise power delivered by the source in bandwidth B).

For real thermal sources, T_{eff} is the actual physical temperature of the source. However, there is a calculated T_{eff} for noise sources, such as atmospheric interference, which are not true thermal noise sources. The method of calculating the T_{eff} for a noise source which does not have a physically measurable temperature is instead to measure the noise power received, N, from the source and then to insert this figure in the equation, $N = kTB$, where T is the required T_{eff}:

$$T_{eff} = N/kB$$

In the case of a noisy amplifier, the gain of the device must be taken into account when determining its effective noise temperature, T_A. The equation for calculating T_A is:

$$T_A = 1/(G_A kB) \times \text{(Amplifier noise power output assuming no noise input)}$$

Figure 8.7 shows a situation where a source of noise temperature T_s feeds a signal of power S_i into an amplifier of available power gain G_A. The amplifier itself has an effective noise temperature T_A. Assuming that the amplifier input

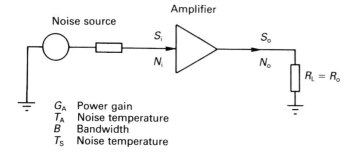

Figure 8.7 *Noise expressed in terms of noise temperatures*

and output resistances are correctly matched by the circuit configuration, we can say that the input noise to the amplifier from the source is:

$$N_i = kT_sB$$

Now, the input SNR is equal to S_i/kT_sB, where B is the bandwidth of the amplifier. Therefore the output power from the amplifier, N_o, is given by (Amplifier gain) × (Input source noise) × (Noise generated in the amplifier), that is:

$$N_i = G_A N_i + G_A kT_A B = G_A kB(T_s + T_A)$$

Now the output signal power $S_o = G_A S_i$ so that the final output SNR is:

$$S_o/N_o = G_A S_i/[G_A kB(T_s + T_A)]$$

and this simplifies to:

$$\frac{S_o}{N_o} = \frac{T_s}{T_s + T_A}\left(\frac{S_i}{N_i}\right)$$

8.10 Noise factor

While noise temperatures are usually used to define the noise characteristics of low noise sources, T_s, and low noise amplifiers, T_A, where T_A is not small when compared with T_s, then the noise factor, F, is used. The noise factor, F, is defined as:

F = (Available noise power at the output)/(Available noise power output assuming that the amplifier itself is noiseless)

This further assumes that the source feeding the noiseless amplifier is at the standard ambient temperature of $T_0 = 290$ K:

$$F = \frac{N_o}{G_A kT_0 B}$$

However, because $N_o = G_A k(T_o + T_A)$ and since $T_s = T_o$,

$$F = \frac{G_A k(T_o + T_o)B}{G_A k T_o B} = \frac{T_o + T_A}{T_o}$$

Transposing this equation gives the following relationship:

$$T_A = (F - 1)T_o$$

Using the noise factor, F, rather than T_A in the expression for the SNR obtained at the end of the preceding section, we can write

$$\frac{S_o}{N_o} = \frac{T_s}{T_s + (F - 1)T_o} \frac{S_i}{N_i}$$

If we now substitute $T_s = T_o$ and rearrange we obtain the result

$$F = \frac{S_i}{N_i} \bigg/ \frac{S_o}{N_o}$$

Noise factors are frequently quoted in decibels. An advantage of expressing noise in this way is that it is then compatible with the term SNR, which is usually expressed in decibels. For example, if the noise factor is 6 dB and the input SNR is 30 dB, then the output SNR, in decibels, is:

$$S_o/N_o = \text{Input SNR (dB)} - \text{Noise factor (dB)}$$
$$= 30 - 6 = 24 \text{ dB}$$

This result can be obtained by the alternative calculation as follows:

10 log $F = 6$ and $F = $ antilog $0.6 = 3.98$

10 log $(S_i/N_i) = 30$

Therefore $S_i/N_i = 10^3$

Now, $S_o/N_o = 1/F \times S_i/N_i = 10^3/3.98 = 251.25$

So, S_o/N_o (dB) $= 10$ log $251.25 = 24$ dB

8.11 Noise factor and temperature of series amplifiers

Figure 8.8 shows two amplifiers in series with a noise source. The noise generated by the noise source is amplified by both amplifiers as is the noise generated by the first amplifier. The noise generated by the second amplifier is amplified only by itself. The following discussion will show that, no matter how many stages of amplification are in series, the system as a whole is only as noisy as its first stage.

Suppose that for the two stages, the respective power gains, noise factors and temperatures are G_1, F_1, T_1 and G_2, F_2 and T_2. Let the system bandwidth be B and the system noise source temperature be T_s.

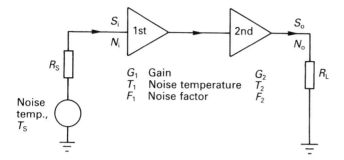

Figure 8.8 *Noise temperature and noise factor considerations in a two-stage amplifier*

The total system noise output, N_o, will be the sum of the noises caused by the (noise source) + (1st amplifier) + (2nd amplifier):

$$N_o = G_1 G_2 (kT_sB) + G_1 G_2 (kT_1B) + G_2(kT_2B)$$
$$= G_1 G_2 kB(T_s + T_1 + T_2/B)$$
$$= G_1 G_2 kB(T_s + T_{eff})$$

where $T_{eff} = T_1 + (T_2 + T_{eff})$. T_{eff} is the effective noise of the system.

This result can be expressed in another way if we use noise factors instead of noise temperatures. The effective noise factor of the system, F_{eff}, is

$$F_{eff} = F_1 + (F_1 - 1)/G_1$$

Because the whole system noise is effectively decided by the level of noise generated in the first stage, it is important that this stage is designed to be as noise free as possible; even to the extent of sacrificing gain.

8.11.1 Problem 1

A data signal source has a noise temperature of 20 K and feeds an amplifier of noise temperature 200 K and power gain 60 dB. The amplifier bandwidth is 20 dB. Assuming matched amplifier input and output resistance conditions, calculate: (a) the noise power at the output; and (b) the output SNR.

Calculations
(a) The noise output power is

$$N_o = GkT_sB + GkT_AB = GkB(T_s + T_A)$$

Now 60 dB gain is the same as $G = 10^6$, $k = 138 \times 10^{-23}$, $T_s = 20\,\mathrm{K}$, $T_A = 200\,\mathrm{K}$ and $B = 1.5 \times 10^6\,\mathrm{Hz}$, so

$$N_o = 10^6 \times 1.38 \times 10^{-23} \times 1.5 \times 10^6 (20 + 200)$$
$$= 4.55 \times 10^{-9}\,\mathrm{W}$$

(b) The output SNR is

$$\frac{S_o}{N_o} = \frac{T_s}{T_s + T_A}\frac{S_i}{N_i}$$

Now $S_i/N_i = 20$ dB is the same as 100 times, so

$$\frac{S_o}{N_o} = \frac{20}{20 + 200} \times 100 = 9.09 \ (= 9.59 \ \text{dB})$$

8.11.2 Problem 2

The block diagram of a low noise data acquisition receiver is shown in Figure 8.9. Calculate: (a) the effective noise temperature and noise factor of the receiver; (b) the available signal power input to the receiver to provide at least a 20 dB output SNR.

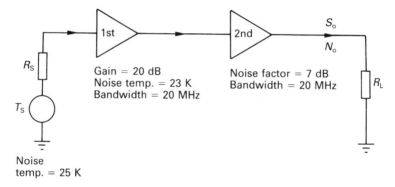

Figure 8.9 *Block diagram for the problem worked in Section 8.11.2*

Calculation

(a) The effective noise temperature of the receiver

$$T_{\text{eff}} = T_1 + \frac{T_2}{G_1}$$

where $T_1 = 23$ K, the noise temperature of the 1st stage; $G_1 = 100$ (being 20 dB), the power gain of the 1st stage; and $T_2 = (F_2 - 1)T_o$, the noise temperature of the 2nd stage.

Now $10 \log F_2 = 7$ dB, so the 2nd stage noise factor is $F_2 = \text{antilog } 0.7 = 5$. $T_o = 290$K, the standard ambient temperature, so $T_2 = (5 - 1)290 = 1160$ K. Therefore

$$T_{\text{eff}} = (23 + 1160/100) = 34.6 \ \text{K}$$

The effective noise factor, $F_{eff} = F_1 + (F_1 - 1)G_1$ where $F_1 = (T_o + T_1)$ $T_o = (290 + 23)/290 = 1.079$, the noise factor of the 1st stage. Therefore the effective noise factor for the receiver is

$$F_{eff} = 1.079 + (5 - 1)100 = 1.119$$

(b) The total noise output, N_o, is given by

$$N_o = G_1 G_2 k B T_{eff} + G_1 G_2 k B T_s$$

The signal power output $= S_o = G_1 G_2 S_i$ and the input noise power $= N_i = k B T_s$. The output SNR is S_o/N_o where

$$\frac{S_o}{N_o} = \frac{G_1 G_2 S_i}{G_1 G_2 k B (T_{eff} + T_s)} = \frac{S_i}{k B (T_{eff} + T_s)}$$

and if S_o/N_o has to be at least 20 dB, then we can say

$$S_i = k B (T_{eff} + T_s) \times (S_o/N_o)$$
$$= 1.38 \times 10^{-23} \times 20 \times 10^6 \times (34.6 + 25) \times 100$$
$$= 1.64 \times 10^{-12} \text{W or } 1.64 \text{ pW}$$

Exercises

8.1 A signal source of 1 mV r.m.s. and internal resistance of 2.7 kΩ at 17°C feeds an amplifier of input resistance 3 kΩ and 1 MHz bandwidth. Calculate the amplifier input SNR due to the thermal noise generated by the source. (Take $k = 1.38 \times 10^{-23}$ J/K.)

8.2 A 30 μW signal, having a SNR of 35 dB, is amplified by an amplifier of power gain 20 dB. The internal noise generated by the amplifier is equivalent to an additional noise power of 6 nW at its input. Calculate:
   (a) the input noise power caused by the 35 dB input SNR;
   (b) the total output noise power;
   (c) the output signal power; and
   (d) the output SNR.

8.3 (a) Calculate the r.m.s. noise voltage generated by a 120 kΩ resistor at 20°C over a bandwidth of 0.5 MHz.
   (b) Calculate the maximum available noise power from a resistance at 20°C between the frequencies of 500 kHz and 2 MHz.
   (c) Calculate the maximum available noise power which can be fed into an amplifier, having a 6 MHz bandwidth, from a source of noise temperature 900 K.
   (d) An amplifier has a noise temperature of 90 K, a power gain of 20 dB and a bandwidth of 200 kHz. Calculate the noise power output caused by the amplifier's own internally generated noise. ($k = 1.38 \times 10^{-23}$ J/K.)

8.4 An amplifier of power gain 35 dB generates an equivalent noise power referred to its input terminals of 50 nW. If a signal power of

-20 dBm and a SNR of 20 dB is applied to the amplifier input, calculate:

(a) the output signal power; and
(b) the output SNR.

8.5 An antenna of noise temperature 160 K feeds a signal to an amplifier of noise temperature 260 K, bandwidth 2 MHz and power gain 90 dB. The SNR at the amplifier input is 30 dB. Assuming k to be 1.38×10^{-23} J/K, calculate the output signal power and the output SNR.

8.6 A 20 μV r.m.s. signal source of internal resistance 60 Ω delivers an input to an impedance matched receiver which has a bandwidth of 1.5 MHz. If the SNR at the receiver output is 10 dB, the ambient temperature is 290 K and Boltzmann's constant is 1.38×10^{-23} J/K, calculate the noise factor of the receiver.

8.7 An antenna, delivering a signal power of 0.3 pW, has a noise temperature of 600 K and is matched to the input of a two-stage amplifier. The first stage has a noise factor of 3 dB and a power gain of 10 dB. The matched second stage has a noise factor of 6 dB and a bandwidth of 10 kHz. Calculate the SNR at the receiver output.

9

Analogue signal filters*

9.1 Introduction

In the context of this chapter, filters are electrical networks that have been designed to pass alternating currents generated at only certain frequencies and to block or attenuate all others. Filters have a wide use in electrical and electronic engineering and are vital elements in many telecommunications and instrumentation systems where the separation of wanted from unwanted signals – including noise – is essential to their success. There are two generic types of filter: passive and active. The first type comprises simple resistors, capacitors and inductors while the second has the addition of active components, usually in the form of operational amplifiers. Both of these types are subdivided into the four classes according to their use. These are low-pass, high-pass, band-pass and band-stop. This chapter is mainly concerned with active filters employing operational amplifiers, but it may serve as a useful introduction for some readers if firstly a brief examination is made of the passive type.

9.2 Passive filters

9.2.1 The low-pass filter

The circuit of a simple CR low-pass filter is shown in Figure 9.1. This is essentially a potential divider comprising a resistance in series with a capacitor. The output voltage, e_o, is taken from across the capacitor and is related to the input voltage, e_i, by the equation:

$$e_o = -jX_c e_i/(R - jX_c)$$

Algebraic manipulation of this complex number equation shows that the amplitude of e_o is given by the expression:

$$|e_o| = e_i X_c/\sqrt{(R^2 + X_c^2)}$$

Even though e_i may be held constant over a range of input frequencies, the amplitude of e_o decreases as the frequency is increased. This is because the reactance of the capacitor, $X_c = 1/2\pi fC$, varies as the inverse of the frequency, f, and tends from an infinitely high value at zero frequency to zero at an infinitely high frequency. The circuit output effectively is shorted out at very high frequencies. Figure 9.2 shows the response curve for this circuit which is typical of the low-pass filter.

* From Clayton and Newby (1992).

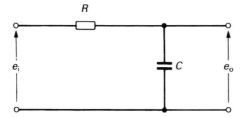

Figure 9.1 *First order, low-pass passive* CR *filter circuit*

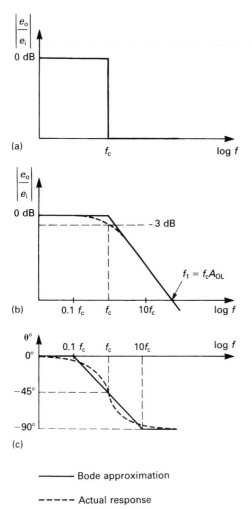

Figure 9.2 *Low-pass filter response curves for: (a) ideal magnitude; (b) actual magnitude with Bode approximation; (c) phase shift with Bode approximation*

At low frequencies the output volts/input volts ratio remains sensibly level up to a frequency, f_c, at which a marked fall off starts. At about $2f_c$ the fall off (or roll-off, as it is usually called) becomes linear at 20 dB per decade (which is the same as 6 dB per octave). The frequency f_c is known as the cut-off frequency and is taken as that frequency at which the reactance of the capacitor has the same magnitude as the resistance in the circuit. Also, f_c is the frequency at which the output voltage has fallen to $1/\sqrt{2}$ times its d.c. value to give half the d.c. power output. Simple calculations based on these facts show that the cut-off frequency is given by the equation:

$$f_c = 1/2\pi RC \text{ (Hz)}$$

For frequencies below f_c the circuit gain (output volts/input volts) is taken as being reasonably constant while for frequencies higher than f_c the gain is regarded as being so low that the passage of these signals is effectively blocked. The circuit is known as a low-pass filter having a bandwidth extending from d.c. to f_c.

Because the response of the circuit depends upon frequency to the mathematical first order, the filter is known as a first order filter. (Also note that the circuit contains only a single component – the capacitor – the performance of which is frequency conscious.)

The ideal low-pass passive filter frequency response curve or transfer function would show no loss of gain for frequencies below f_c and zero output above f_c (see Figure 9.2). Clearly, the first order low-pass filter achieves neither of these ideals. If two *CR* sections are cascaded (see Figure 9.3) to form a second order filter having two frequency dependent capacitors, a steeper roll-off can be obtained, but only at the expense of decreased output. If these two similar sections are used, the roll-off tends to 40 dB per decade but the output is so attenuated as to be of little use.

A better solution for achieving a steep roll-off is still to use two frequency dependent components but make one a capacitor and the other an inductance. This circuit, shown in Figure 9.4, takes advantage of the ability of the inductance and capacitance to be near their natural resonant frequency at the filter cut-off frequency. This would have the effect of producing an output voltage magnification in the knee region of the frequency response curve. By varying the ratio of the values of inductance and capacitance, the shape of the knee can be adjusted. The critical case is where the flat top of the lower frequency response is extended along the frequency scale before falling in a steeper roll-off yet without introducing the undesirable effects of underdamping or overdamping. These include output voltage oscillations before finally settling or having an excessively long response time to transient inputs. The combined high frequency effect of the high inductive reactance coupled with the low capacitive reactance eventually produces a second order filter linear roll-off dependent upon the inverse square of the frequency.

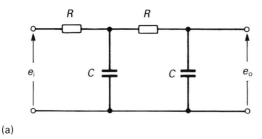

(a)

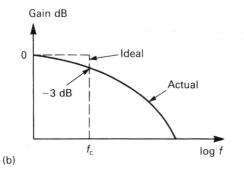

(b)

Figure 9.3 *Second order, high-pass passive* CR *filter circuit (a) and response curve (b)*

9.2.2 The high-pass filter

To form a high-pass filter, the CR components of the low-pass filter are simply interchanged. Figure 9.5 shows the first order high-pass circuit and Figure 9.6 its frequency response curve. The gain roll-off is once again 20 dB per decade and the cut-off frequency is still given by the equation:

$$f_c = 1/2\pi RC \text{ (Hz)}$$

At low input frequencies the capacitor has a high reactance and effectively rejects any input voltage. As the input frequency is increased the capacitor progressively lowers its reactance, allowing an increasing proportion of the input voltage to be developed across the resistor and appear at the circuit output. Frequencies below f_c are regarded as being in a stop-band; those above, as being in the circuit pass-band.

9.2.3 The band-pass filter

A second order band-pass filter can be obtained by using the series LCR circuit arrangement shown in Figure 9.7. At low input frequencies the capacitive reactance predominates and the circuit behaves as a simple series capacitor with a 6 dB per octave increasing response from d.c. As the frequency of the

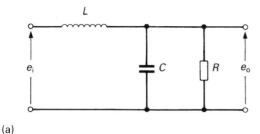

(a)

(b)

Figure 9.4 *Second order, low-pass passive* **LCR** *filter circuit (a) and response (b)*

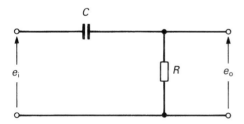

Figure 9.5 *First order, high-pass passive* **CR** *filter circuit*

input signal approaches circuit resonance, there is a marked upturn in the response curve to climax in a peak at:

$$F_o = 1/2\pi\sqrt{(LC)} \text{ (Hz)}$$

Once the resonant frequency has been exceeded, the inductive reactance becomes increasingly dominant and the response falls away but not as sharply as was the build up from the low frequencies.

Thus there are two frequencies where the response is 3 dB less than the peak and they are called the upper and lower cut-off frequencies, f_{cu} and f_{cl}. They are

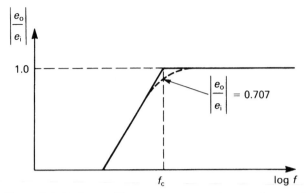

Figure 9.6 *High-pass filter response curve showing the actual (dotted) response and the Bode approximation*

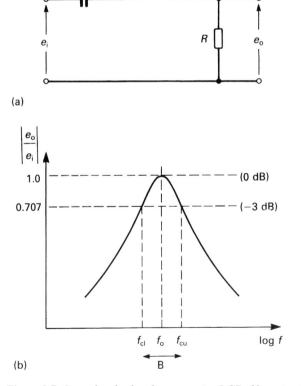

(a)

(b)

Figure 9.7 *Second order band-pass passive* LCR *filter circuit (a) and response (b)*

not equally disposed about the resonant or centre frequency; the centre frequency is always taken as the geometric mean of the two:

$$f_0 = \sqrt{(f_{cu} f_{cl})} \ (\text{Hz})$$

The difference between f_{cu} and f_{cl} is taken as the bandwidth or pass-band, B, of the filter and together with the goodness or Q of the circuit is related to the centre frequency, f_0, by the following equation:

$$B = f_0/Q \ (\text{Hz})$$

The higher the Q of the circuit the smaller is its pass-band and the filter is said to be more selective.

If several CLR circuits, each having a slightly different resonant frequency, are connected in series, the resulting circuit is a band-pass filter.

9.2.4 The band-stop filter

A second order band-stop filter can be obtained by using the parallel LC circuit arrangement shown in Figure 9.8. At low input frequencies, the circuit is effectively a low-pass arrangement comprising only the L and the R. At the circuit resonant frequency, determined by $f = 1/2\pi\sqrt{(LC)}$, the parallel L and C presents an infinitely high impedance and the circuit output is zero. Once the resonant frequency has been exceeded, the inductive reactance continues to increase while that of the capacitor decreases, making the circuit perform more as if comprising only the C and the R in a simple high-pass filter arrangement.

9.2.5 Passive filter summary

The four basic frequency sensitive filter circuits described above can be cascaded using any mix of first and second order variants that is necessary to produce a desired response. The shape of filter response curves has been studied by many eminent people, some of whom have had their names credited to particular circuits which satisfy particular requirements. These names include Butterworth, Bessel, Chebyshev and Cauer, and, together with other special filter circuits, they will be discussed later in this chapter.

Passive filter circuits contain various combinations of resistors, capacitors and inductors and in most cases suffer from several shortcomings. Mathematically, they are difficult to design; they are often pulled off frequency by the load current drawn from them; even in their pass-band they usually attenuate signals and are not easily tuned over a wide frequency range without changing their response characteristics. Further problems can be associated with the use of inductors. Not only are they expensive, bulky and heavy, but they are also prone to magnetic field radiation unless expensive shielding is used to prevent unwanted coupling.

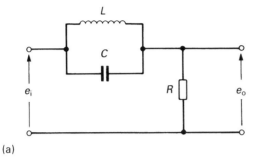

(a)

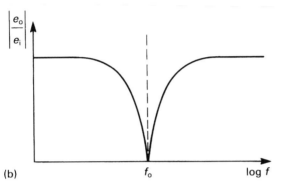

(b)

Figure 9.8 *Second order band-stop passive* LCR *filter circuit (a) and response (b)*

9.3 Active filters

9.3.1 The case for active filters

The advent of the low cost, integrated circuit operational amplifier has made it possible to overcome most of the problems associated with the passive filter circuit. Not only will the high input impedance and low output impedance of the operational amplifier effectively isolate the frequency sensitive filter network from the following load, but it can also provide useful current or voltage gain. More significantly, the operational amplifier can be designed into a *CR* only circuit in such a way as to provide a filter response virtually identical with that of a passive inductive filter network. This means that the use of inductors in filters now is unnecessary. Unlike the inductor, the operational amplifier does not possess a magnetic field which stores energy; rather it is designed to behave mathematically in the same way as the whole passive circuit it replaces. Any additional circuit energy is obtained from the separate power source used by the operational amplifier.

9.3.2 Negative impedance conversion

The circuit shown in Figure 9.9 is designed to have an input impedance, Z_i, which appears to be the negative of the impedance Z.

$$i_i = (e_X - e_o)/R \qquad (9.1)$$

Normal operational amplifier action causes

$$e_i = e_X = e_Y \text{ and } e_Y = e_o Z/(R + Z)$$

Rearranging,

$$e_o = e_1(R + Z)/Z \qquad (9.2)$$

Substituting (9.2) and $e_X = e_i$ into (9.1),

$$i_i = \frac{e_i - \dfrac{e_i(R + Z)}{Z}}{R} = \frac{e_i\left(1 - \dfrac{R + Z}{Z}\right)}{R}$$

Therefore

$$i_i = \frac{e_i(Z - R + Z)}{ZR}$$

and rearranging further,

$$Z_i = \frac{e_i}{i_i} = \frac{ZR}{Z - R - Z}$$

or

$$Z_i = -Z$$

Suppose Z is a capacitor, C. Then $Z = -j/\omega C$ and so it follows that:

$$Z_i = -(-j/\omega C) = +j/\omega C$$

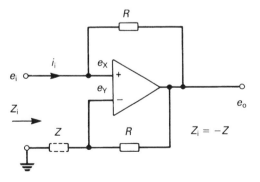

Figure 9.9 *Impedance converter*

The $+\mathrm{j}$ means that the current lags the voltage, that is, has an inductive reactance, $\mathrm{j}X_{\mathrm{L}}$, but where $X_{\mathrm{L}} = 1/\omega C$. However, while the result is to produce an inductive effect, the 'inductive reactance' decreases with increasing frequency rather than increases as would the reactance of a true inductance.

9.3.3 Impedance gyration

A single negative impedance inverter is not capable of simulating the true action of an inductor. However, this effect can be achieved if a pair of negative impedance converters is used. Such a circuit is shown in Figure 9.10 where the input impedance is $Z_{\mathrm{i}} = R^2/Z$. Suppose that $Z = -\mathrm{j}X_{\mathrm{c}}$, then

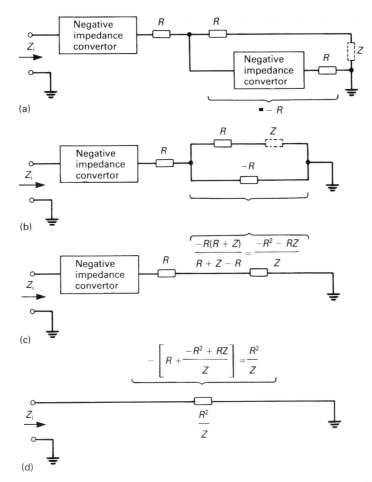

Figure 9.10 *Analysis of the circuit in (a) by progressive simplification through (b) and (c) to (d) shows that the input impedance is proportional to the reciprocal of the load impedance,* Z

$Z_i = R^2 / -jX_c = +j\omega CR^2$. Now the capacitor C is being made to act as if it were a true inductor of value $L = CR^2$.

Similarly, it can be shown that if Z were an inductive reactance of value jX_L, then the gyrator would make this appear to the preceding circuit as a true capacitance of $C = L/R^2$.

9.3.4 Making a simple active filter

The response curve shown in Figure 9.4 for a passive second order low-pass RL filter can be simulated using only resistors and capacitors. A first attempt may include two cascaded first order, low-pass CR sections with the addition of an emitter follower. This has a high input impedance but low output impedance and so minimises any loading effects on the frequency sensitive CR sections. This circuit and the highly damped response that it produces are shown in Figure 9.11(a).

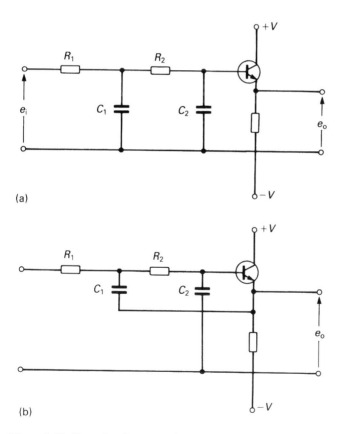

Figure 9.11 *Second order active filter*

Figure 9.11(b) illustrates a major design improvement by the introduction of positive energy feedback to the centre of the CR section. This 'bootstrapping' has a maximum effect only near the cut-off frequency. At very low frequencies the normal gain enhancement of positive energy feedback is largely negated by the high reactance of C_1 in the feedback path. In excess of the cut-off frequency, the low reactance of C_2 allows the signal to leak to earth and attenuate the output accordingly. The values of the filter network capacitors and resistors can be selected to eliminate the damping problem of the previous circuit. While the product of the resistors and capacitors decides the cut-off frequency, it is the ratio of the capacitors which affects the circuit response rate. Compared with the values of C_1 and C_2 for critical damping, a large C_1 with a small C_2 will produce an underdamped response while the reverse will cause overdamping.

9.4 Active filters using operational amplifiers

In practical active filters, the emitter followers used above invariably are replaced by operational amplifiers in the form of integrated circuits. The frequency sensitive filtering networks are placed either before the operational amplifier input terminals or in the feedback circuits.

9.4.1 First order high-pass and low-pass filters

Examples of simple first order high- and low-pass active filters are shown in Figure 9.12. As expected, the frequency selective resistor–capacitor circuit elements decide the frequency response. The cut-off frequency is $f_c = 1/2\pi CR$, at which the magnitude of the filter response is 3 dB less than that in the pass-band, and the higher frequency roll-off tends to 20 dB per decade. If a low value of f_c is required, a general purpose bi-FET operational amplifier should be suitable. This will allow the use of large resistance values without introducing any appreciable bias current offset error. Resistor values up to 10 MΩ may be used, so avoiding the expense of a high value, close tolerance capacitor.

First order low-pass filters are often used to perform a running average of a signal having high frequency fluctuations superimposed upon a relatively slow mean variation; for this purpose it is simply necessary to make the filter time constant, CR, much greater than the period of the high frequency fluctuations.

A practical point to remember is that all operational amplifier active high-pass filters show a band-pass characteristic. This is because their response eventually falls at frequencies which exceed the closed loop bandwidth of the operational amplifier.

9.4.2 Second order low-pass and high-pass filters

Examples of simple second order low-pass and high-pass active filter circuits are shown in Figure 9.13. The second order filter response has a 40 dB per decade roll-off in the stop-band. The sharpness of the response curve knee

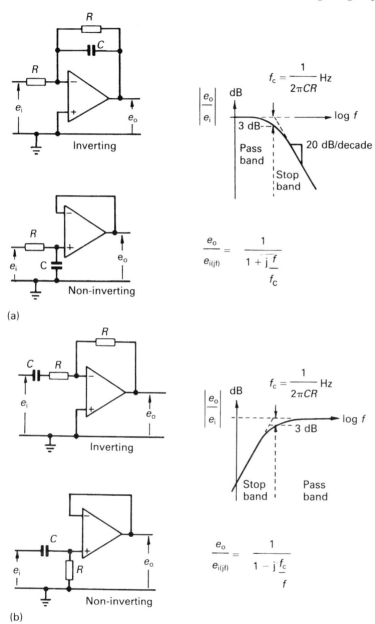

Figure 9.12 *First order low- and high-pass active filters. (a) First order low-pass response. (b) First order high-pass response*

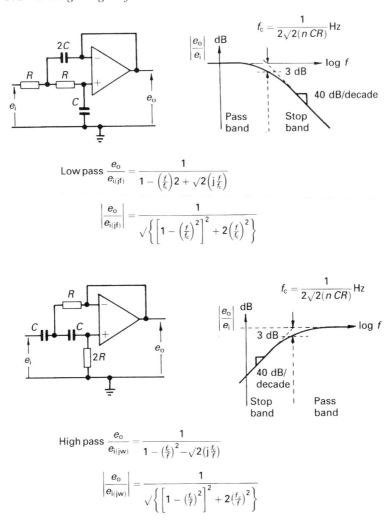

$$f_c = \frac{1}{2\sqrt{2}(n\,CR)}\,\text{Hz}$$

Low pass $\dfrac{e_o}{e_{i(jf)}} = \dfrac{1}{1 - \left(\frac{f}{f_c}\right)2 + \sqrt{2}\left(j\frac{f}{f_c}\right)}$

$$\left|\frac{e_o}{e_{i(jf)}}\right| = \frac{1}{\sqrt{\left\{\left[1 - \left(\frac{f}{f_c}\right)^2\right]^2 + 2\left(\frac{f}{f_c}\right)^2\right\}}}$$

$$f_c = \frac{1}{2\sqrt{2}(n\,CR)}\,\text{Hz}$$

High pass $\dfrac{e_o}{e_{i(jw)}} = \dfrac{1}{1 - \left(\frac{f_c}{f}\right)^2 - \sqrt{2}\left(j\frac{f_c}{f}\right)}$

$$\left|\frac{e_o}{e_{i(jw)}}\right| = \frac{1}{\sqrt{\left\{\left[1 - \left(\frac{f_c}{f}\right)^2\right]^2 + 2\left(\frac{f_c}{f}\right)^2\right\}}}$$

Figure 9.13 *Second order low- and high-pass active filters*

depends upon the choice of values for the components forming the frequency sensitive element of the filter. In Figure 9.13, the components are proportioned to give a so-called Butterworth response (see Section 9.5.1) and the cut-off frequency $f_c = 1/[2\sqrt{2}(\pi CR)]$ Hz.

9.5 Choosing the frequency response of the low-pass filter

Figure 9.2 shows the ideal shape for a low-pass filter. It has a perfectly flat (horizontal) response from zero frequency up to the cut-off frequency where a vertical fall then occurs. In practice this perfectly rectangular shape is unattainable. Depending upon the intended role of the filter, it can be designed to approximate to the ideal response in varying ways and these are described briefly below.

9.5.1 The Butterworth low-pass response

This response requires that at zero frequency the circuit gain is flat and remains as near flat as possible up to the designed cut-off frequency. The higher the order of filter the more accurately does its response approximate to this ideal, as illustrated in Figure 9.14.

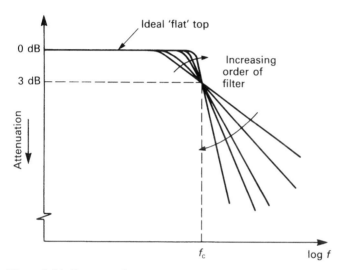

Figure 9.14 *Butterworth response*

9.5.2 The Chebyshev low-pass response

The Chebyshev approximation is an attempt to overcome the practical failure of the Butterworth response to maintain a truly flat pass-band as the frequency of operation is increased up to the cut-off frequency. The Chebyshev circuit is designed uniformly to spread any deviation of gain over the pass-band in the form of ripples as shown in Figure 9.15. Above the cut-off frequency, like the Butterworth response, the Chebyshev roll-off eventually tends to be monotonic at $20n$ dB per decade where n is the order of the filter. Even so, the second and third order Chebyshev filters tend to have a less steep initial roll-off than their Butterworth counterparts whereas comparable fourth order and above filters show the Chebyshev response to have the sharper knee.

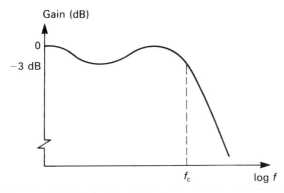

Figure 9.15 *Third order Chebyshev response*

9.5.3 The Cauer (or elliptic) low-pass response

Using a Butterworth or Chebyshev filter, a complete signal stop is usually regarded as having been achieved when the filter attenuation has reached a designed level. The frequency at which this degree of attenuation first occurs is taken as the start of the filter stop-band. However, while a continued increase in frequency initially causes further signal attenuation, a practical limit is reached. This is where, because of unwanted leakage through stray reactances, further increasing the frequency can produce an unwanted output from the filter. The Cauer response is designed to cater for those applications where it is required that an infinite attenuation is achieved at a particular frequency and that for any higher frequencies a designed minimum attenuation is maintained.

Figure 9.16 shows the Cauer filter circuit diagram and the typical response curves it produces. The infinite attenuation is caused at the frequency, f_2, because at this frequency L_2 and C_2 are in resonance and present an infinite impedance to the signal flow.

9.5.4 The Bessel low-pass response

The above studies on the Butterworth, Chebyshev and Cauer filter responses have all emphasised the relative amplitudes of the filter input and output voltages. No mention has been made of the phase shift which occurs as the signal travels through the filter. In applications involving voice or other analogue transmissions, phase shift is not important and optimum amplitude responses are often obtained at the expense of phase shift. However, in the case of digital transmissions it can be important that the pulses are not distorted and linear-phase filters are often used.

Figure 9.17 shows the ideal linear relationship between the signal frequency and the resulting phase shift introduced by the filter. With regard to the signal transit time through the filter, ideally, signals of all frequencies should suffer the same time delay and so any signals in phase at the input will still be in phase

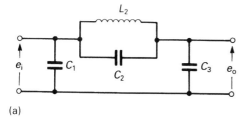

(a)

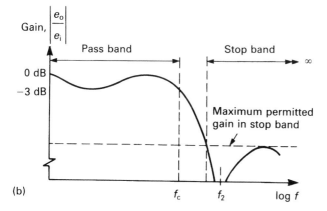

(b)

Figure 9.16 *Cauer or elliptic, third order filter circuit (a) and the response it produces (b)*

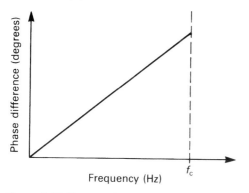

Figure 9.17 *Linear relationship between frequency and phase shift in ideal low-pass filter*

at the output. But a signal of double the frequency of another will suffer twice its phase shift. This effect is shown in Figure 9.18.

The Bessel approximation is an attempt to produce such a linear-phase filter. The Bessel response circuit has the same appearance as the Butterworth and Chebyshev circuits and differs only in the component values necessary to produce the required constant transit time at all frequencies.

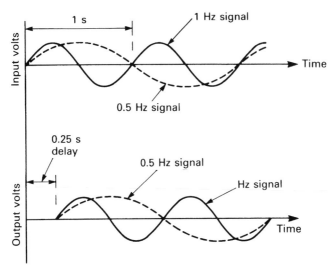

Figure 9.18 *Time related waveforms showing how a constant time delay of 0.25 s produces a 90° phase shift in a 1 Hz signal but only a 45° shift in a 0.5 Hz signal*

9.5.5 Comparative responses of the different low-pass filters

See Figure 9.19 for a summary of the various low-pass responses.

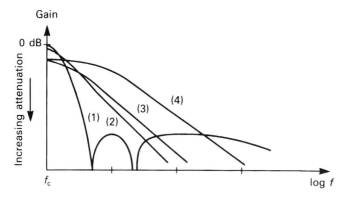

Figure 9.19 *Comparison of the different low-pass filter performance in their stop bands. (1) Cauer (elliptic); (2) Chebyshev; (3) Butterworth; (4) Bessel*

9.6 Choosing the frequency response of the high-pass filter

Figure 9.6 shows the ideal shape for a high-pass filter response curve. It has a zero output at low frequencies but continued frequency increase eventually causes the response to rise monotonically until just short of the −3 dB cut-off frequency which marks the start of the pass-band. At frequencies higher than this, the response in the pass-band levels at the maximum gain. But, because of practical component inadequacies and stray reactances becoming increasingly significant at the higher frequencies, the flat response of the passive filter circuit element does not extend to infinity and eventually declines. Additionally, in the case of the active filter, the inherent high frequency gain roll-off of the operational amplifier effectively makes any high-pass filter behave as a form of band-pass filter − but with the upper cut-off frequency being above the highest frequency to be passed.

The studies of the low-pass filtering transfer functions and response curves made in Section 9.5 readily can be modified to suit the high-pass conditions. Basically, the high-pass filter is a mirror image of its low-pass equivalent; the capacitors and resistors are simply interchanged. The mathematical process involved in this change is called mathematical transformation by $1/f$. Figure 9.20 shows a graphical summary of the high-pass response curves.

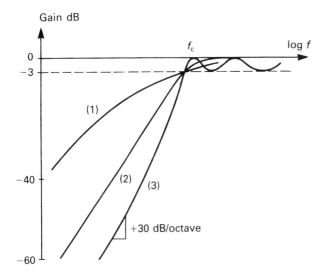

Figure 9.20 *Example of high-pass filter (fifth order) response curves. (1) Overdamped; (2) flattest; (3) Chebyshev*

9.7 Band-pass filters using the state variable technique

A band-pass filter can be constructed by connecting a high- and a low-pass filter in series. This method may suffice for some applications but if a very narrow pass-band is required then we require a high Q characteristic and a different filter construction is required. Because high Q filters using only a single operational amplifier tend to be sensitive to variations in the values of the circuit components, three operational amplifiers are used in the state variable technique. Today, with inexpensive quad operational amplifier chips being readily available, the use of several operational amplifiers in a single filter circuit is no problem.

Figure 9.21 shows a second order state variable filter. Not only will this circuit produce a pass-band output, but it can also provide a high-pass and a low-pass output at the same time. The addition of a fourth amplifier (not shown) can even produce a band-stop or 'reject' effect. The derivation of the transfer characteristics takes into account the integrating action of amplifiers A_2 and A_3. Recall that the action of an integrator is effectively to multiply the input by $-1/j\omega T$ where $T = CR$, the integration time constant.

Amplifier A_1 adds the input signal to the output of amplifier A_2 and subtracts the output of amplifier A_3. The output of amplifier A_1, e_{A1}, is therefore

$$
e_{A1} = \frac{e_{bp}}{-\dfrac{1}{j\omega T_1}}
$$

$$
= \left(e_i \frac{R_4}{R_3 + R_4} + \frac{e_{bp} R_3}{R_3 + R_4} \right)\left(1 + \frac{R_6}{R_5} \right) - \left(-\frac{1}{j\omega T_2} e_{bp} \frac{R_6}{R_5} \right)
$$

where $T_1 = C_1 R_1$ and $T_2 = C_2 R_2$.
This can be shown to reduce to

$$
\frac{e_{bp}}{e_{i(j\omega)}} = \frac{-\dfrac{1}{T_1}\dfrac{\left(1 + \dfrac{R_6}{R_5}\right)}{\left(1 + \dfrac{R_3}{R_4}\right)} j\omega}{\dfrac{1}{T_1 T_2}\dfrac{R_6}{R_5} + j\dfrac{\omega}{T_1}\dfrac{\left(1 + \dfrac{R_6}{R_5}\right)}{\left(1 + \dfrac{R_4}{R_3}\right)} - \omega^2}
\tag{9.3}
$$

This can be written in the form of a second order band-pass transfer function as follows:

$$
\frac{e_{bp}}{e_{i(j\omega)}} = -\frac{A_{obp}}{1 + jQ\left(\dfrac{\omega}{\omega_o} - \dfrac{\omega_o}{\omega}\right)}
\tag{9.4}
$$

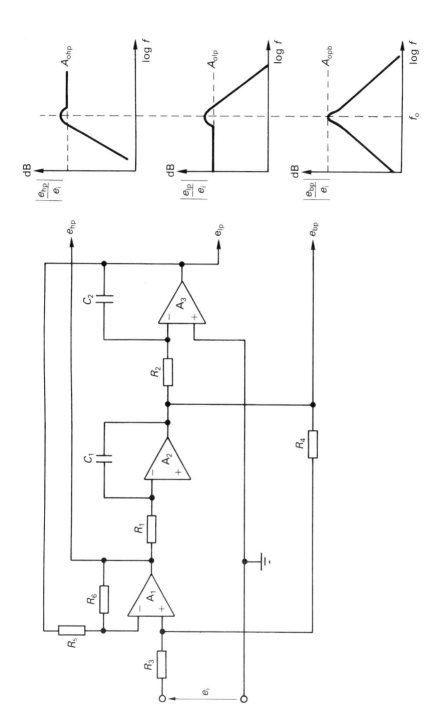

Figure 9.21 *State variable filter technique*

The constants in Equation 9.4 are related to the circuit component values as:

$$A_{\text{obp}} = \frac{R_4}{R_3}$$

$$\omega_0 = \sqrt{\frac{R_6}{R_5 C_1 R_1 C_2 R_2}}$$

$$Q = \sqrt{\frac{C_1 R_1 R_6}{C_2 R_2 R_6}} \; \frac{1 + \dfrac{R_4}{R_3}}{1 + \dfrac{R_6}{R_5}}$$

(9.5)

If we make $R_5 = R_6$, $C_1 = C_2$ and $R_1 = R_2$, the centre frequency, ω_0, can be changed by the simultaneous adjustment of R_1 and R_2. This does not affect the value of Q, which can be adjusted without affecting ω_0 by changing R_4.

The low-pass and high-pass response can be ascertained by the respective substitution into Equation 9.3 of the following terms:

$$e_{\text{lp}} = -(1/j\omega T_2) e_{\text{bp}} \quad \text{and} \quad e_{\text{hp}} = -(j\omega T_1) e_{\text{bp}}$$

The second order low-pass response becomes

$$\frac{e_{\text{lp}}}{e_{i(j\omega)}} = \frac{A_{\text{olp}}}{1 + 2\zeta j \dfrac{\omega}{\omega_0} - \left(\dfrac{\omega}{\omega_0}\right)^2}$$

(9.6)

The constants in the equation are given by the following relationships:

$$A_{\text{olp}} = \frac{1 + \dfrac{R_5}{R_6}}{1 + \dfrac{R_3}{R_4}}$$

$$\omega_0 = \sqrt{\frac{R_6}{R_5 C_1 R_1 C_2 R_2}}$$

$$\zeta = 0.5 \frac{1 + \dfrac{R_6}{R_5}}{1 + \dfrac{R_4}{R_3}} \sqrt{\frac{C_2 R_2 R_5}{C_1 R_1 R_6}}$$

where ζ is the damping factor.

The second order high-pass response becomes

$$\frac{e_{\text{hp}}}{e_{i(j\omega)}} = \frac{A_{\text{ohp}}}{1 - 2\zeta j \dfrac{\omega_0}{\omega} - \left(\dfrac{\omega_0}{\omega}\right)^2}$$

(9.7)

The constants in the equation are given by:

$$A_{\text{olp}} = \frac{1 + \dfrac{R_5}{R_6}}{1 + \dfrac{R_3}{R_4}}$$

$$\omega_o = \sqrt{\frac{R_6}{R_5 C_1 R_1 C_2 R_2}}$$

$$\zeta = 0.5 \frac{1 + \dfrac{R_6}{R_5}}{1 + \dfrac{R_4}{R_3}} \sqrt{\frac{C_2 R_2 R_5}{C_1 R_1 R_6}}$$

9.8 Filter design

The preceding paragraphs have given an insight into the different shapes of filter response curve which may be obtained by the careful selection of the order of filter required together with the correct component values. The mathematical prediction of a particular response using manual methods becomes lengthy, tedious and error prone as the filter order increases. The recent proliferation of personal computers has made these design calculations a less onerous task, but even more important is the availability of ready-made designs for which tables of 'normalised' frequency against component values have been published and which can be used to design a filter having a particular cut-off frequency and input impedance. The tabulated 'normalised' figures are 'scaled' to give practical component values.

9.8.1 Normalisation and scaling

Suppose we consider one of the simplest active filter circuits, the single pole low-pass filter, a typical circuit for which is shown in Figure 9.22(a). (The feedback resistor is included for d.c. offset purposes.) It is shown in Section 9.2.1 that the cut-off frequency for this CR circuit is given by:

$$f_c = 1/2\pi CR \text{ (Hz) or } \omega_c = 1/CR \text{ rad/s}$$

If the circuit were required to have an impedance level of 1 Ω and a cut-off angular frequency of 1 rad/s, then the capacitor would need a value of 1 F. The circuit would be said to have been 'normalised' to 1 Ω and 1 rad/s and is shown in Figure 9.22(b). With these values for resistance and capacitance the circuit is not of great use but it can be 'scaled' to determine the values of resistance and capacitance that give a particular cut-off frequency and impedance level.

Suppose we wish to raise the impedance level from the normalised 1 Ω to 500 Ω. The rule for this is to raise all the circuit impedances by a factor of 500. This means that all the resistances must be multiplied by 500 but that all the

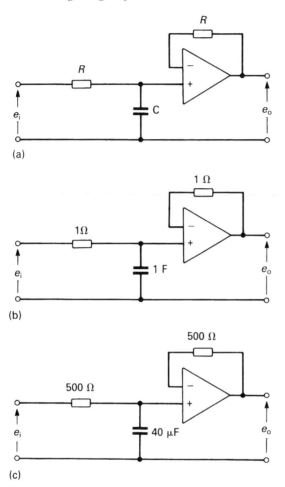

Figure 9.22 *Simple first order low-pass active filter. (a) Typical circuit. (b) Circuit normalised to 1 Ω impedance level and cut-off frequency of 1 rad/s. (c) Circuit scaled to change the impedance level to 500 Ω and cut-off frequency to 50 rad/s (7.96 Hz)*

capacitances will require their values to be divided by 500, since capacitive reactance is inversely proportional to frequency.

Rule 1   To increase the impedance level multiply the resistors and divide the capacitors by the scaling factor.

Further suppose that we now wish to increase the cut-off frequency from 1 rad/s to 50 rad/s without changing the newly adjusted impedance level. The requirement now is for the circuit time constant to be reduced in that same ratio, that is by 1/50. This means that the product of CR must be reduced by

1/50 without altering the fixed value of R at 500 Ω. Therefore, from the relationship $\omega = 1/CR$ and knowing that ω must be 50 rad/s and that R is newly fixed at 500 Ω, C becomes $1/(50 \times 500)$, which is 40 F.

Rule 2 To increase the cut-off frequency, divide either the resistors or the capacitors by the scaling factor.

When the circuits comprise more than a single resistor with a single capacitor, as is the case with the higher order filters, the same basic rules still apply. But remember, for multisection filters, the ratios of the frequency sensitive capacitor and resistor pairs must remain unchanged if the overall filter frequency response is not to change. Also, if the frequency of one section is altered then all sections must be changed to the same frequency.

It was shown in Figure 9.12(a) that the closed loop gain for this single pole low-pass filter is:

$$\frac{e_o}{E_{i(jf)}} = \frac{1}{(1 + f/f_c)} = \frac{1}{\left(1 + \dfrac{\omega}{\omega_c}\right)}$$

where $e_{i(jf)}$ simply means that e_i is a complex number dependent upon frequency.

The gain can be expressed in polar form as follows:

$$\frac{e_o}{e_i} = \frac{1}{\sqrt{1 + \left(\dfrac{\omega}{\omega_c}\right)^2} \angle \tan^{-1} \dfrac{\omega}{\omega_c}}$$

$$\frac{e_o}{e_i} = \frac{1}{\sqrt{1 + \left(\dfrac{\omega}{\omega_c}\right)^2}} \angle - \tan^{-1} \dfrac{\omega}{\omega_c}$$

This equation can be used to calculate the circuit gain for varying values of ω, these being made a known fraction of ω_c. The table of data together with the plotted response curve are shown in Figure 9.23.

9.8.2 Sallen–Key second order active filters

There are many circuit configurations which operate successfully as second order filters, but perhaps any dissertation on active filters would be incomplete without at least a mention of the circuits jointly attributable to Sallen and Key. There are two basic Sallen–Key designs: the unity gain filter and the equal-component filter.

While these circuits are relatively simple to construct, in order that they operate as expected, the various component values must have a definite relationship which is a function of the circuit Q factor. Typical Sallen–Key second order low-pass filter circuits of the two types mentioned are shown in

ω (rad/s)	$\left|\dfrac{e_o}{e_i}\right|$	Phase angle (degrees)
0.0625 ω_c	0.998	− 3.58
0.100 ω_c	0.995	− 5.71
0.125 ω_c	0.992	− 7.13
0.25 ω_c	0.970	− 14.0
0.5 ω_c	0.894	− 26.6
ω_c	0.707	− 45.0
2 ω_c	0.447	− 63.4
4 ω_c	0.243	− 76.0
8 ω_c	0.124	− 82.9
10 ω_c	0.100	− 84.3
16 ω_c	0.062	− 86.4
100 ω_c	0.010	− 89.4

(a)

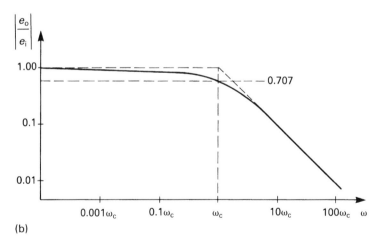

(b)

Figure 9.23 *Response curve and data table for a first order low-pass filter. (a) Table of data for closed loop magnitude and phase. (b) Response curve plotted from data in (a)*

Figure 9.24. It is important to note that both these circuits have constant gains; one being unity, the other 2:1.

An advantage of the unity gain circuit is that it requires the minimum number of components; even the feedback resistor is not necessary in some circuits and may be omitted. However, the unity gain circuit does not lend itself to easy conversion to a high-pass or band-pass filter by simply interchanging the circuit positions of some components. On the other hand, the equal-component circuit requires more components but has the advantage of being simple to convert to a high-pass filter by interchanging the frequency determining resistors and capacitors.

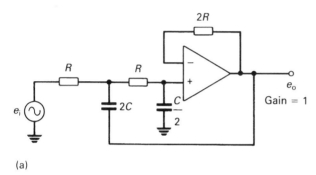

(a)

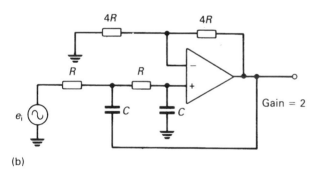

(b)

Figure 9.24 *Second order Sallen–Key, low-pass active filters. (a) Typical unity gain circuit; (b) typical equal-component circuit*

Exercises

9.1 Show that for the circuit in Figure 9.1 the cut-off frequency, f_c, is given by the equation $f_c = 1/2\pi RC$ (Hz).

9.2 Also for the circuit in Figure 9.1, show that the magnitude of the voltage output, $|e_o|$, and its phase angle, ϕ, are given, respectively, by the equations: $|e_o| = e_i X_c/\sqrt{R_2 + X_c^2}$ and $\phi = \tan^{-1}(X_c R)$.

9.3 Plot a graph of gain, $|e_o/e_i|$, against frequency for a simple first order, passive, low-pass, *CR* filter and use this graph to show that:
   (a) at f_c the gain is 3 dB less than its d.c. value; and
   (b) at frequencies higher than $2f_c$ the output declines at 6 dB/octave.

9.4 A second order, low-pass filter with a Butterworth response and a 3 dB cut-off frequency of 20 Hz is required. The filter circuit in Figure 9.13 is to be used with close tolerance capacitors of value 0.001 μF and 0.002 μF.
   Calculate:
   (a) the values of resistors that are required; and
   (b) the frequency at which the filter gain will have fallen to half its d.c. value.

9.5 It is decided to use the circuit shown in Figure 9.13 as a high-pass, active filter having a cut-off frequency of 1125 Hz. The capacitors are both to be 0.001 μF. Calculate the values of the resistors to meet this design specification.

9.6 If the high-pass, active filter circuit in Figure 9.13 has a 3 dB cut-off frequency of 1125 Hz and its input is connected to a 500 mV variable frequency signal, estimate the voltage output from the filter at: (a) 100 Hz, (b) 1 kHz, and (c) 2 kHz.

9.7 It is required to produce a unity gain, low-pass Sallen–Key active filter based on the circuit shown in Figure 9.24(a). If the filter input resistance is to be 20 kΩ and the 3 dB cut-off frequency 1 kHz, suggest suitable values for the capacitors and resistors.

9.8 Component values $R_3 = R_5 = R_6 = 15$ kΩ, $C_1 = C_2 = 0.01$ μF, $R_1 = R_2 = R_4 = 500$ kΩ are used in the state variable filter shown in Figure 9.21.
   (a) Calculate the constants A_{obp}, f_o and Q for the amplifier A_2.
   (b) Sketch the response curve showing the gain in decibels against log f.

10
Modulation and demodulation

10.1 The need for modulation

Instrumentation systems usually involve some degree of signal conditioning and if the data to be conditioned are to be transmitted over line or radio links then this may be best done using a form of *modulation*. Suppose, for example, a process system has several types of sensor for the monitoring of temperature, pressure, angular velocity, position and so on. These quantities are first transduced into electrical analogue signals before being sent to a signal conditioning centre. After being suitably conditioned, the signals may be required to be transmitted back to the original process system to undertake controlling adjustments. Of course, the signals representing the different quantities need to be kept separate, or confusion will ensue. Therefore, the signals representing each quantity are transmitted along their own special signal channel using a system of modulated carrier waves.

10.1.1 Line communications

If the distance between the process system and the signal conditioning equipment is short, it may well be economical to use a pair of lines to carry each signal channel. However, should the number of signal channels be large and the transmission distance long, then it may be more economical to use only one pair of lines and separate the different signals electronically rather than physically. This can be done using one of two techniques known as time division multiplexing (t.d.m.) or frequency division multiplexing (f.d.m.).

10.1.2 Radio communications

The distance the instrumentation data have to be transmitted, both before and after conditioning, may be such that a radio link must be used. This is typical of satellite control for example. As with line communications, the same problem exists, that is the need to transmit many data channels simultaneously without their interfering one with another. With radio a second problem arises. This is that low frequency signals cannot easily be transmitted as electromagnetic radio waves. The main reason why this is so is that in order to transmit a radio wave efficiently, the aerial length must be one-quarter that of the signal wavelength. For very low frequencies, for example 800 Hz, the aerial would need to be approximately 96 km long. This is not a practical proposition so a shorter aerial must be used. But this reduces the efficiency of the transmission by requiring a disproportionately large amount of power to obtain any meaningful radiation. In practice, for frequencies below 50 kHz, the aerial limitations make radio links impracticable. The answer would appear to be the translation of the necessarily low frequency (d.c. in many cases) sensor analogue signals into frequencies higher than 50 kHz. This is therefore a further call for *modulation*. The action of modulating a signal is to impress upon that signal a second signal. The first signal is of a higher frequency than the second. The first signal is called the *carrier* signal and the second the *modulating* signal. The modulating signal is the one which contains the instrumentation data or other useful information. The sending end of a data link requires a modulator to impress the data onto a carrier, while at the receiving end a demodulator is required to extract the modulation or data from the received modulated carrier. Figure 10.1 shows the block diagram and waveforms of a simple data transmission link using a modulated carrier.

In Figure 10.1(a), the sending end modulator is fed from a data acquisition system with, for example, positive and negative d.c. pulses, A, which represent some physical quantity. In the modulator, these data pulses are impressed upon a high frequency sinusoidal carrier voltage, B, which is also fed into the modulator from a constant frequency oscillator. A is impressed upon B in the manner indicated by the waveform sketches in Figure 10.1(b). The modulated carrier wave emanating from the modulator is a signal of constant frequency B but whose *amplitude* is varying at a rate decided by the data pulse, A, the overall result being the modulated carrier wave C. The positive data pulses increase the carrier amplitude while the negative ones make it decrease. After transmission over a line or radio link, the data pulses, D, which are a copy of A, are extracted from the modulated carrier by a technique called *demodulation* which we shall examine later.

The transmission of several modulated carrier channels over a line or radio link is often achieved using the above modulation process. The separation of the different data channels using either line or radio links can be arranged by *frequency division* or *time division multiplexing* techniques. We shall now look at each of these in turn.

10.2 Frequency division multiplexing (f.d.m.)

The action of transmitting several simultaneous signals over a single communications link is called *multiplexing*. Using f.d.m., each data channel is allocated a unique carrier frequency. The separation of the different data channels has to recognise the fact that a modulated carrier wave contains more frequencies than simply the carrier frequency. In the simplest case, the modulating signal is a sine wave of frequency f_m and this is made to modulate (or change) the amplitude of a carrier sine wave of frequency f_c. The resulting modulated carrier will effectively comprise three sine waves. These are the original carrier frequency, f_c, of the original carrier amplitude, and two more of frequency $f_c - f_m$ and $f_c + f_m$, respectively, each of half the original modulating signal amplitude. The components of the modulated signal can be represented diagrammatically as shown in Figure 10.2(a). The two sine waves at frequencies $f_c - f_m$ and $f_c + f_m$ are called *side frequencies*. The carrier frequency and the side frequencies are each represented by vertical arrows on a horizontal

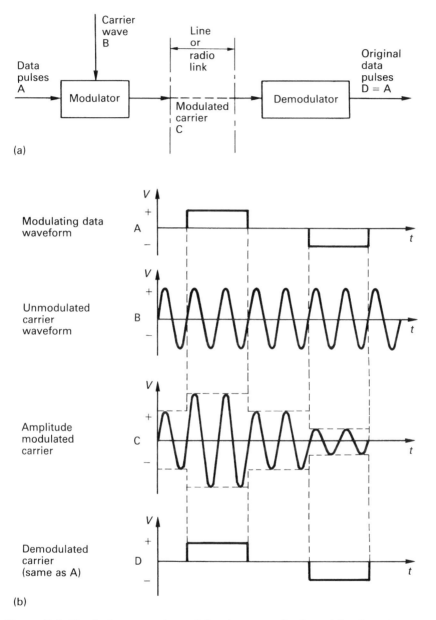

Figure 10.1 *Simple data transmission link using an amplitude modulated carrier wave.*
(a) Block diagram; (b) waveforms

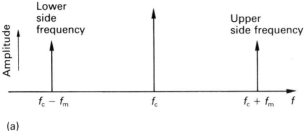

(a)

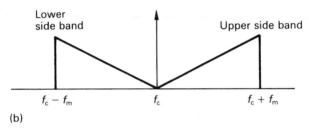

(b)

Figure 10.2 *Diagrammatic representation of an amplitude modulated carrier wave. (a) Carrier at frequency* f_c *modulated by a single wave at frequency* f_m. *(b) Carrier at frequency* f_c *modulated by several waves having a maximum frequency* f_m

frequency axis. Should the modulating signal consist of more than a single frequency then a whole series or band of side frequencies result. The two side frequencies become the upper and lower side bands and are represented by triangles the ordinate heights of which are proportional to the modulating frequency. If the highest modulating frequency is f_m, then the upper side band extends up to a frequency of $f_c + f_m$ and the lower side band down to $f_c - f_m$. The two side bands are mirror images of each other about f_c as the centre frequency. They carry the useful information being transmitted by the modulated carrier wave.

10.2.1 Carrier suppression

In an amplitude modulated carrier wave, the useful information being transmitted is all contained in the side bands. But a large proportion of the power in the modulated carrier wave is in the carrier itself. Therefore, in order to reduce the transmission of unnecessary power, the carrier frequency can be removed or *suppressed*. This reduces the power handling requirement of the transmission equipment, which can be solely concerned with the information frequencies in the side bands. Complete suppression of the carrier in radio transmissions is rare because of the technical difficulties which arise in the demodulation of the suppressed carrier signals.

10.2.2 Side band suppression

Since the two side bands contain the same information, there is no requirement for both to be transmitted. This means that the data can be transmitted by the suppression of the carrier and one of the two side bands. If this is done an immediate advantage is the reduction in bandwidth requirement of the transmitted signal. With modern communications ever crowding the available frequency spectrum this is an attractive feature. It is normal practice in carrier communication to wholly or partially suppress one of the side bands, depending upon the nature of the transmitted signal. The saving in bandwidth is illustrated in Figure 10.3. The suppression of the carrier and side bands is undertaken by the use of electronic filters. The techniques of filtering are covered in some detail in Chapter 9.

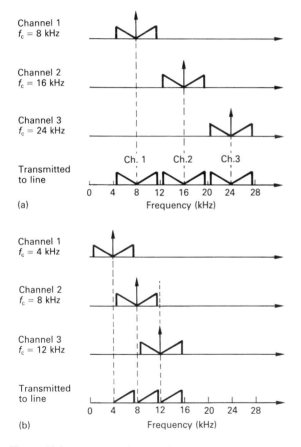

Figure 10.3 *Diagrams showing double and single side band working. (a) Double side band system, carrier suppressed. (b) Single side band system carrier and lower side band suppressed*

10.3 Time division multiplexing (t.d.m.)

In this system, the different communication channels are separated by time rather than by frequency. There is a single transmission path which is time-shared between the number of channels wishing to use it. The principle involved is shown in Figure 10.4. Here five channels each take a turn in being routed through the common line or radio link. Thus, each channel transmits only a succession of samples of its sending end signal rather than its true continuous version. At the receiving end, a copy of the original continuous signal is reconstructed from these samples.

The two rotary switches, SWA and SWB, in Figure 10.4 are continually stepping round in synchronism. Therefore, each channel has the exclusive use of the transmission path for the duration of the periods that the wipers touch the particular channel contacts. While Figure 10.4 depicts the rotary switches as being mechanical devices, the modern versions use solid state electronic switches. The logic diagram of an electronic equivalent of the rotary switch system is shown in Figure 10.5. The send and receive gates are all shown as AND gates which are activated or *enabled* in sequence by a system of timing or *clocking* pulses. These are applied simultaneously to one input terminal on each of the relevant channel gates at the send and receive ends. With the respective data channels continuously presenting their signals on the other input terminals of the send gates, the application of a clock pulse establishes the communication data link for the duration of the clock pulse. The four clock pulses for the four channel system depicted in Figure 10.5 are time staggered so that only one pair of gates is opened at any one time.

The time interval during which a particular channel is open is called its *time slot*. Its width will depend upon the total number of channels being handled by the multiplexing system and the recurrence frequency of the clock pulses. This latter is known as the *sampling frequency*. For example, suppose that 24 channels are being multiplexed and that the whole group is being sampled 8000 times per second. If the slot width is δ, we can say $\delta = T/24$, where T is the sampling frequency period. $T = 1/8000$ s, therefore, $\delta = 1/(24 \times 8000)$ s $=$ 5.2 μs.

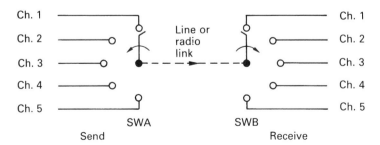

Figure 10.4 *Time division multiplexing*

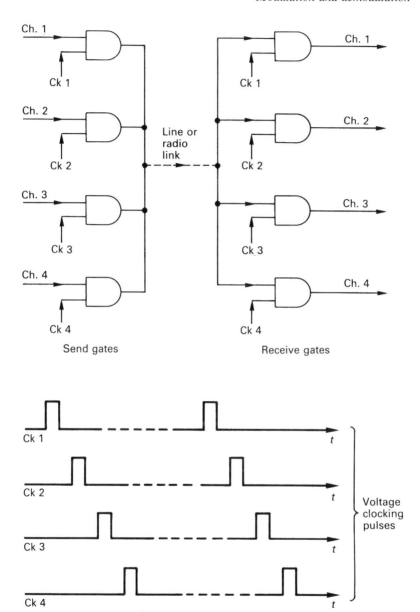

Figure 10.5 *The logic diagram equivalent of the rotary switch system shown in Figure 10.4*

The channel signal on the common link between the send and receive ends of the t.d.m. system will comprise a series of pulses the heights of which depict the amplitudes of the modulating data signals at the instants of sampling. This situation is shown in Figure 10.6. For the sake of simplicity, only two channels of a group of four channels is shown; two have been omitted. The information contained by the modulating data is assumed to be sinusoidal and the two channels, 3 and 4, are shown as solid and broken lines respectively. The regular sampling times, or channel time slots, for the four channels are shown together with the sample of pulse information produced. The information is contained in the height of the pulse and is independent of the pulse width.

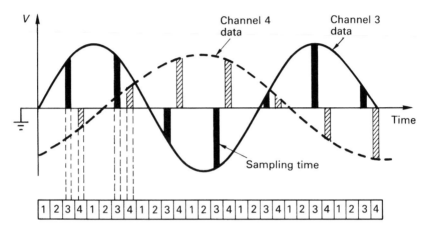

Time when channels are connected

Figure 10.6 *Timing diagram showing time sampling for sinusoidal data on channels 3 and 4*

10.3.1 Pulse amplitude modulation (p.a.m.)

A simple system which could produce the multiplexed p.a.m. of the circuit in Figure 10.6 need be no more than a simple switch connecting a data or modulating signal generator, f_m, to a resistor across which the data pulse is generated. This is shown in Figure 10.7(a). Figure 10.7(b) shows the closing and opening times of the switch and Figure 10.7(c) shows the instantaneous amplitude, e_1, of the modulating signal being sampled by the switch action. Figure 10.7(d) is the resulting p.a.m. signal, e_2, comprising the required sample pulses.

The disadvantage of the p.a.m. system is that the required information is contained in the height of each individual pulse. If, in the course of transmitting these pulses down a transmission link, the pulses were to become distorted owing to attenuation and phase shifting, then the received information may not truly represent that originally sent. The pulses when reconstituted into the analogue waveform may be distorted outside acceptable

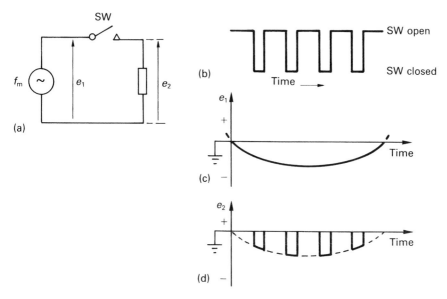

Figure 10.7 *Pulse amplitude modulation. (a) The circuit. (b) Switch timing diagram. (c) Sample of e_i. (d) Pulse amplitude modulated representation of (c)*

limits. This is quite a serious disadvantage; so much so that an alternative system of pulse modulation was introduced which is not so sensitive to amplitude distortion. This is the pulse code modulation (p.c.m.) system.

10.3.2 Pulse code modulation

This is a development of the p.a.m. system described above. The sample pulses at the send end are examined for their height. Each pulse, according to its height, is then translated into a group of constant amplitude pulses arranged in a coded pattern. The data signal to be transmitted has thus been changed into one of constant height pulses, the height and width of which do not convey information. The information is conveyed only by the number and pattern of the transmitted pulses. Much distortion of the transmitted pulses can now be tolerated because at the receive end the problem of reading the pulses is simply one of detecting their presence or absence.

The process of converting the height of the p.a.m. pulses into binary code is called *quantisation*. Figure 10.8 helps to illustrate the mechanics of the quantisation process. The original data voltage waveform is sampled at regular intervals, so producing a train of amplitude modulated pulses (p.a.m.). The heights of these pulses are individually measured and each value given a binary code. This is done by dividing the analogue data waveform variations into amplitude levels as shown in Figure 10.8. These levels are called quantum levels. The measured pulse height is allocated to the nearest quantum level each of which has a unique binary code of four binary digits (bits). For the 4-bit

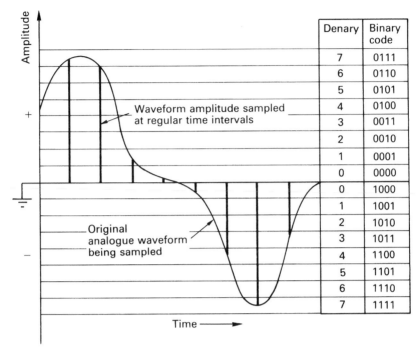

Figure 10.8 *Diagram showing method of coding positive and negative amplitudes*

coding system depicted, there are eight levels (0000–0111) for the positive pulses and a further eight (1000–1111) for the negative pulses. Of the four digits comprising each group, the first (most significant bit (m.s.b.)) digit denotes the sign (+ or −) and the subsequent three digits, the magnitude or quantum level. In this particular case, positive numbers are prefixed by a '0' and negative numbers by a '1'.

10.3.3 Encoding and decoding

It has been shown above that the amplitude of p.a.m. pulses can be translated into a binary code. The equipment which undertakes this translation from amplitude to binary code is called an *encoder* and the reconstitution of the different pulse amplitudes from the binary code is performed by a *decoder*. Figure 10.9 shows the block diagram of a t.d.m. system using p.c.m. The encoder, which examines each of the pulse magnitudes as they are presented to it by rotary switch SWA, compares the sample with a number of standard magnitudes or quantising levels. It then allocates to the sample the binary code of the nearest quantising level. The resulting stream of constant height pulses, which constitute the coded p.a.m., are transmitted from the encoder over the common data link to the decoder at the receive end. The decoder reconstitutes

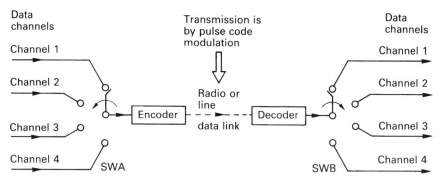

Figure 10.9 *Pulse code modulation data transmission system*

the p.a.m. for subsequent routing through rotary switch SWB to the appropriate data channels.

In one practical system in use, the encoder requires 128 quantising stages between the positive and negative peaks of an analogue signal. The representation of these 128 stages requires seven digits (from $2^7 = 128$). The encoder uses six digits to define the magnitude of the analogue signal and prefixes these with a seventh, 'sign', digit which prescribes whether the signal is positive or negative. Finally, there is an eighth bit which is required for system control purposes. The time order for the transmission of these eight pulses is: control signalling, sign and then the six for amplitude description.

10.3.4 Bandwidth requirements

Figure 10.10 shows the arrangement of the time slots, in a 24 channel system, for a period $1/f_s$ where f_s is the sampling frequency. If f_s is 8 kHz, the time period during which every channel in the system must be sampled once is 125 μs. At each sampling instant, 8 bits are transmitted serially for each channel. Thus, the bit rate per sampling cycle or frame can be found as follows. Because there are 24 channels in the system, the number of bits, n, transmitted in 125 μs is $24 \times 8 = 192$.

Therefore, the bit rate is

$$n = \frac{192}{125 \times 10^{-6}} = 1.536 \times 10^6 \text{ bits/s}$$

The minimum bandwidth required to transmit at this rate is 1.536 MHz.

10.3.5 Unipolar and bipolar transmission

The problem with transmitting a stream of pulses of like polarity is that they produce a fluctuating d.c. potential which varies in magnitude according to the energy storage capacity of the transmission line used. This can cause technical problems if the transmission line is long.

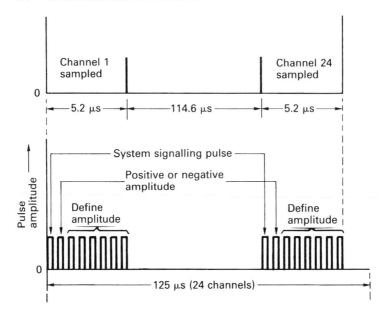

Figure 10.10 *Timing diagrams for first 'look' or sampling of 24 data channels*

Figure 10.11 shows one method used to overcome the disadvantages of this unidirectional or unipolar switching of voltage pulses. The normal binary coded p.a.m. is put through a process (step 1) known as 'alternate digit inversion' followed by a second process (step 2), 'alternate mark inversion'. The end result is a stream of coded pulses which produce an overall zero d.c. charging effect on the transmission line.

10.4 Analysis of an amplitude modulated wave

The mathematical analysis of any modulating system is usually undertaken assuming that the unmodulated carrier and the modulating signal are both sinusoidal. This being the case we can represent the instantaneous value of a sine wave by the equation:

$$a = A \sin(\omega t + \phi)$$

where A is the amplitude or peak value, ω is the angular frequency, t is the elapsed time, and ϕ is the phase angle.

Modulation of the sinusoidal carrier wave can be undertaken in several ways. One is to modulate or vary A (amplitude modulation); another is to modulate ω (frequency modulation); and yet another is to modulate ϕ (phase modulation). At this juncture we are to consider amplitude modulation only but we shall look at frequency and phase modulation later.

A sinusoidal carrier wave whose amplitude is modulated by a second sinusoidal signal may be considered to be either: a carrier wave whose

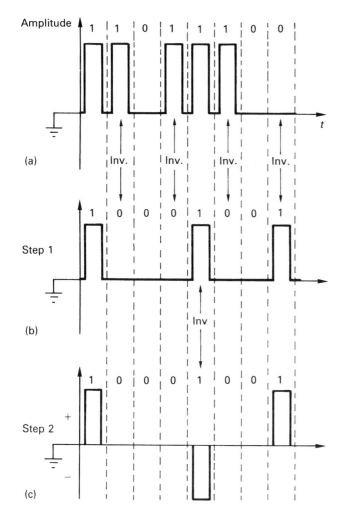

Figure 10.11 *Binary code transformation into an alternate mark inversion form. (a) Normal binary code of a p.a.m. sample. (b) Changes alternate digits after first digit. (c) Inverts alternate +1 to −1 after first +1*

amplitude varies about its nominal value at the frequency of the modulating signal; or as the nominal carrier wave plus two other constant amplitude sine waves, respectively above and below the carrier frequency by the modulating frequency. These latter two frequencies are known as the upper and lower side frequencies.

The modulated carrier wave as would appear on an oscilloscope trace is depicted in Figure 10.12(c) and is the result of combining (a) and (b) in a modulator. The variation in the carrier peak values where the modulation

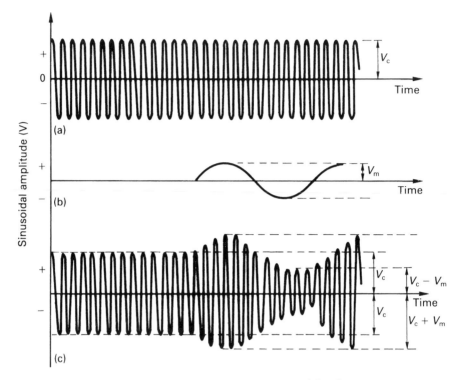

Figure 10.12 *Analysis of amplitude modulation. (a) Unmodulated carrier wave at frequency ω rad/s. (b) Modulating wave at frequency ρ rad/s. (c) Modulated carrier wave at frequency ω rad/s*

occurs can be seen clearly. The modulated carrier positive peaks are made to vary between amplitude limits of $+ (V_c + V_m)$ and $+ (V_c - V_m)$. At the same time a similar variation occurs to the negative peaks of the carrier and the combined result is the *modulation envelope* within which the modulated carrier amplitude excursions are contained. This modulated waveform can be regarded as containing the three constant amplitude sine waves as shown in Figure 10.13.

Having a closer look at the mathematics involved we can see that the instantaneous value of the unmodulated carrier wave is

$$v_c = V_c \sin \omega t$$

and that of the modulating wave is

$$v_m = V_m \sin \rho t$$

If the peak value of the carrier wave is made to vary in sympathy with the modulating wave, then its value is given by the expression $V_c + V_m \sin \rho t$. The equation for the instantaneous value of the modulated carrier voltage, v_{mc}, is

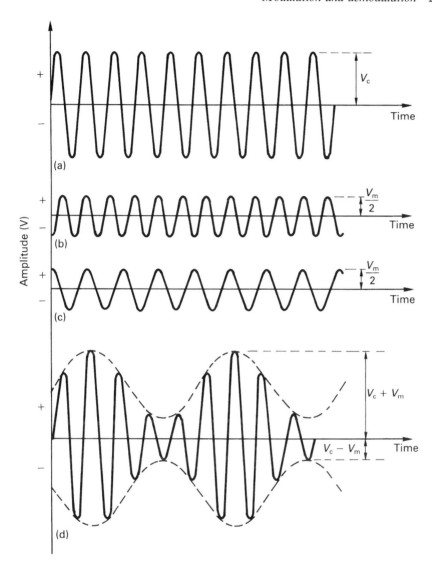

Figure 10.13 *Amplitude modulated carrier can be regarded as comprising three separate frequencies. (a) Carrier wave. (b) Upper side frequency. (c) Lower side frequency. (d) Modulated carrier wave*

$$v_{mc} = (V_c + V_m \sin \rho t) \sin \omega t \ (V)$$
$$= V_c \sin \omega t + V_m \sin \omega t \sin \rho t \ (V)$$

This can be expanded using the identity

$$2 \sin A \sin B = \cos(A - B) - \cos(A + B)$$

to obtain

$$v_{mc} = V_c \sin \omega t + (V_m/2)[\cos(\omega - \rho)t - \cos(\omega + \rho)t] \ (V)$$
$$= V_c \sin \omega t + (V_m/2) \sin[(\omega - \rho)t + \pi/2] + \qquad (10.1)$$
$$(V_m/2) \sin[(\omega + \rho)t - \pi/2] \ (V)$$

The first term in this equation is that of the original unmodulated carrier while the other two terms represent the two side frequencies. The modulated carrier wave is thus shown to comprise the three separate sine waves. The amplitudes of the side frequencies are not the same as that of the original modulating wave, V_m; only half, that is $V_m/2$. The two side frequencies are evenly disposed about the main carrier frequency and separated from it by ρ rad/s. The three constituent sine waves are shown on a frequency spectrum diagram in Figure 10.14. The figure shows the situation where a 10 kHz carrier wave is being modulated by a 1 kHz signal. The lower and upper side frequencies so produced are shown as having only half the modulating signal amplitude and spaced by 1 kHz from the carrier, at 9 kHz and 11 kHz respectively.

If more than one signal frequency is used to modulate the carrier wave, then a series of side frequencies will result. In fact, two side bands will be formed, an upper and a lower side band. The bandwidth of each of the two side bands will be from the lowest modulating frequency to the highest. The overall bandwidth of the two side bands including the carrier will be twice the highest modulating frequency (see Figure 10.2).

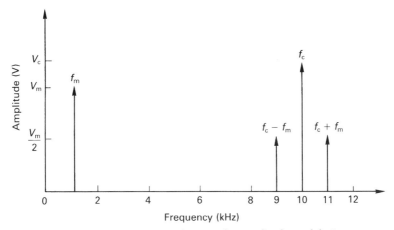

Figure 10.14 *Frequency spectrum diagram for amplitude modulation*

Example 10.1

A data signal comprises a band of frequencies from 20 Hz to 4.75 kHz. It is required that the data be transmitted over a radio link using amplitude modulation and a carrier frequency of 500 kHz. Calculate the bandwidth of the transmission and the highest and lowest frequencies in the modulated carrier wave.

The bandwidth will be twice the highest modulating frequency. Therefore the bandwidth will be

$2 \times 4.75 \text{ kHz} = 9.5 \text{ kHz}$

The upper side band extends from

$500 \text{ kHz} + 20 \text{ Hz} = 500.02 \text{ kHz}$

to

$500 \text{ kHz} + 4.75 \text{ kHz} = 504.75 \text{ kHz}$

The lower side band extends from

$500 \text{ kHz} - 4.75 \text{ kHz} = 495.25 \text{ kHz}$

to

$500 \text{ kHz} - 20 \text{ Hz} = 499.98 \text{ kHz}$

The complete transmitted signal will contain frequencies in the ranges *500.02 kHz to 504.75 kHz and 495.25 kHz to 499.98 kHz*. Also included will be the carrier frequency itself, *500 kHz*. (There will be no frequencies between 499.98 kHz and 500 kHz nor between 500 kHz and 500.02 kHz.)

10.5 Modulation factor and percentage modulation

Figure 10.12(c) of the amplitude modulated carrier wave shows the peak value of the carrier to be V_c while that of the modulating wave is V_m. For the sinusoidal waveforms that we are considering, the *modulation factor*, *m*, is effectively

$m = V_m/V_c$

The expression for the instantaneous modulated carrier voltage can incorporate the term, *m*, by rearrangement as follows:

$$V_{mc} = (V_c + V_m \sin \rho t) \sin \omega t$$
$$= V_c[1 + (V_m/V_c) \sin \rho t] \sin \omega t$$
$$= V_c(1 + m \sin \rho t) \sin \omega t$$

A similar exercise can be undertaken with the instantaneous currents involved. If we call the instantaneous carrier current i_c, where $i_c = I_c \sin \omega t$, and the instantaneous modulating current i_m, where $i_m = I_m \sin \rho t$, then the instantaneous modulated carrier current, i_{mc}, can be shown to be:

$$i_{mc} = I_c(1 + m \sin \rho t) \sin \omega t$$

The *percentage modulation* is the modulation factor expressed as a percentage:

$$\text{Percentage modulation} = (V_m/V_c) \times 100\%$$

If V_m is a quarter of the amplitude of V_c, then the percentage modulation is 25%. The carrier wave is said to be modulated to a depth of 25%. Figure 10.15 illustrates this point, showing (a) 50% and (b) 100% modulation.

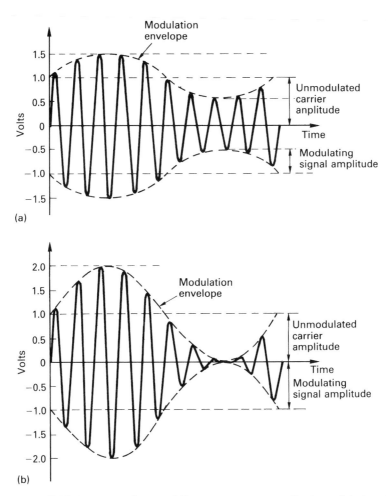

Figure 10.15 *Diagrams showing different percentage amplitude modulation: (a) 50%; (b) 100%*

The amplitudes of the side frequencies produced by different depths of modulation can be determined by inspection of the diagrams in Figure 10.15. For example, for the 50% modulation depth case, V_m can be seen to be varying between 0.5 V and 1.5 V; its amplitude is 0.5 V. We saw earlier that the side frequency amplitude is a half that of the modulating waveform amplitude. So we conclude that the side frequency amplitude is a half of 0.5 V, i.e. 0.25 V. In the case of 100% modulation, the side frequency amplitude is 0.5 V.

A point to note is that more than 100% modulation can be achieved by making the modulating signal of greater amplitude than the carrier itself. This is not good practice since it results in distortion; about 75% maximum for modulation depth is a good guide.

Example 10.2

A high frequency sinusoidal voltage is amplitude modulated by a second lower frequency sinusoidally varying voltage. The resulting modulated carrier varies in amplitude from 1 V to 10 V. Find the modulation factor.

Assuming symmetrical modulation, the swing of the modulated carrier is from 1 V to 10 V, i.e. by 9 V. Therefore, V_c must lie midway between its maximum and minimum excursions at $(1 + 9/2) = 5.5$ V. Now V_m is half the swing of the modulated carrier from peak to trough, i.e. half of 9 V being 4.5 V.

The modulation factor is $V_m/V_c = 4.5/5.5$ or 0.818.

The percentage modulation is 81.8%.

10.6 The power distribution in an a.m. wave

If we take the case of a sinusoidal voltage carrier wave being modulated by another sinusoidal voltage, the result, as shown earlier, is a modulated voltage carrier wave represented by Equation 10.1. This equation is effectively that of three separate sine waves of amplitude V_c, $V_m/2$ and $V_m/2$. If this complex modulated carrier were to be applied across a resistance, R, the total power dissipated in the resistance would be the sum of that produced by each of the three constituent components. This can be done quite easily using the r.m.s. values and remembering that power, P, is $(\text{voltage})^2/R$.

$$P = (V_c/R\sqrt{2})^2 R + (V_m/2R\sqrt{2})^2 R + (V_m/2R\sqrt{2})R$$

By rearrangement and remembering that $V_m/V_c = m$, we can write

$$P = (V_c^2/2R)\,(1 + m^2/2)\ (\text{W}) \tag{10.2}$$

If there is no modulation present, $m = 0$ and the power in the wave is shown to be all in the carrier:

$$P = V_c^2/2R\ (\text{W})$$

If there is 100% modulation of the carrier, $V_c = V_m$, $m = 1$, and

$$P = (V_c^2/2R) (1 + 1/2)$$

Power = Carrier power × 1.5

This means that with 100% modulation, of all the power being transmitted two-thirds is in the carrier and only one-sixth in each of the two side frequencies. The useful information is all contained in one side frequency (or side band); hence the case for transmitting only one side band.

10.7 Frequency modulation

The equation for the instantaneous value of a sine wave is $a = A \sin(\omega t + \phi)$ where $\omega t = 2\pi f_c = \theta$. Thus far we have concerned ourselves with amplitude modulation where information is impressed upon the carrier wave by the variation of A. However, there is another method of modulating a signal. This is called *angle modulation* which can itself be subdivided into: *frequency modulation*, concerned with the variation of θ, and *phase modulation* which is the variation ϕ. For the moment we shall be concerned only with frequency modulation.

A formal definition of frequency modulation is that it is that modulation of a sinusoidal carrier wave in which the instantaneous frequency of the modulated wave differs from the carrier frequency by an amount proportional to the instantaneous amplitude of the modulating wave. The number of times per second that the instantaneous frequency of the modulated wave varies about the carrier centre or nominal frequency is equal to the modulating frequency. More than this, the amplitude of the modulated carrier is kept constant at its unmodulated level. The effect of the modulating signal on the carrier is shown in Figure 10.16.

The way that Figure 10.16 has been drawn shows that the frequency of the carrier wave is increased from its nominal value by increases in the amplitude of the modulating signal. This is not always the case; it can be arranged that increases in modulation amplitude cause the carrier frequency to decrease.

Figure 10.17 is a further representation of the frequency modulation process. Suppose that a 1 kHz sinusoidal signal of peak value A is used to frequency modulate a carrier wave of 500 kHz. The instantaneous frequency of the modulated wave will swing about the unmodulated carrier frequency of 500 kHz, 1000 times per second. The degree of the frequency swing will be proportional to the value of A. If we assume that the frequency swing so caused is 500 Hz above and 500 kHz below the unmodulated carrier frequency, f_c, the instantaneous frequency of the modulated wave will vary from a low of 499.5 kHz to a high of 500.5 kHz.

Doubling the frequency of the modulating voltage will simply double the rate of carrier frequency swing about about its unmodulated centre value, f_c. Doubling the amplitude of the modulating voltage will double the amount the carrier frequency swings away from f_c.

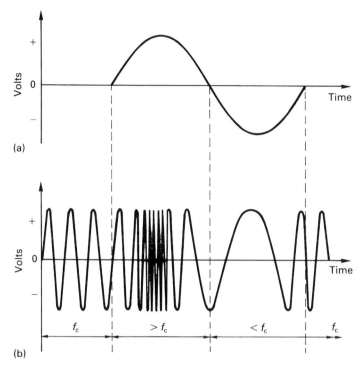

Figure 10.16 *Diagrams showing the frequency variations in a frequency modulated carrier wave. (a) Modulating signal, f$_m$; (b) frequency modulated carrier*

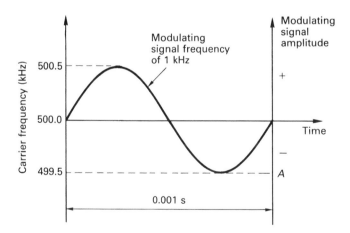

Figure 10.17 *Diagram showing how the instantaneous frequency of the carrier varies with the applied modulation signal. Nominal carrier frequency, f$_c$ = 500 KHz. Modulating signal frequency, f$_m$ = 1 kHz. Carrier frequency change is ±500 Hz*

10.7.1 Frequency modulation terminology

Frequency swing   The modulation action causes the frequency of the unmodulated carrier to alternately increase and decrease. The difference between the highest and lowest instantaneous frequency is known as the frequency swing.

Frequency deviation (f_d)   The frequency deviation is the peak difference between the instantaneous frequency of a frequency modulated wave and the carrier frequency in a cycle of modulation. Frequency deviation is proportional to the amplitude of the modulating signal and may therefore be increased by increasing the amplitude of the modulating signal.

Modulation index (η)   Where the modulating voltage waveform is a sine wave, the modulation index is defined as the ratio of the frequency deviation to the frequency of the modulating waveform:

$$\eta = f_d/f_m$$

Rated system deviation (f_{dr})   This term applies to a system using frequency modulation. It is the maximum frequency deviation undertaken by the carrier wave.

Rated maximum modulation frequency (f_{mr})   This is the maximum modulating frequency that a system can handle.

Deviation ratio (D)   This is the ratio of the rated system deviation (f_{dr}) to the rated maximum modulating frequency (f_{mr}):

$$D = f_{dr}/f_{mr}$$

10.7.2 The mathematical expression of a frequency modulated wave

The expression for a sine wave of amplitude A is $A \sin \theta$, where A is the length of the rotating generating vector and θ is its angular displacement from a given start point in time.

If we denote time as t and the angular velocity of the rotating vector as ω, the graphics for the generation of the sine wave are as shown in Figure 10.18. The instantaneous angular velocity, ω, is defined as the instantaneous rate of change of the angular displacement. Expressed mathematically this is

$$\omega = d\theta/dt$$

If the frequency is held constant, the angular velocity, ω, is also constant and is given by

$$\omega = 2\pi f$$

where ω is in radians per second and f is in hertz.

Therefore the angular displacement, θ_1, after a time interval, t_1, is given by

$$\theta = \omega t_1$$

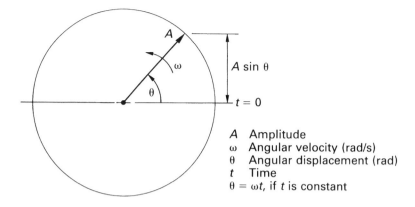

Figure 10.18 *The graphics for the expression* A *sin* θ

The angular displacement is alternatively called the *phase displacement*. Therefore,

$$A \sin \theta_1 = A \sin \omega t_1$$

If the frequency is made to vary with time, then so will the angular velocity. To determine the angular displacement, θ, we must integrate $\omega = d\theta/dt$ with respect to time. Therefore,

$$\theta = \int \omega \, dt$$

But, since $\omega = 2\pi f$, we can say

$$\theta = 2\pi \int f \, dt \qquad (10.3)$$

When a carrier wave, $V_c \sin \theta$, is frequency modulated, the frequency of the modulated carrier wave varies above and below the carrier centre frequency, f_c, by an amount proportional to the instantaneous amplitude of the modulating signal, $V_m \sin \rho t$. The instantaneous frequency of a frequency modulated carrier wave is therefore given by

$$f = f_c + K V_m \sin \rho t \qquad (10.4)$$

where K is a constant of proportionality.

Because $\sin \rho t$ must always lie between -1 and $+1$, the maximum value of $K V_m \sin \rho t$ will be $\pm K V_m$. Therefore, the frequencies of a frequency modulated wave will vary between the limits given by $f_c \pm K V_m$. The peak variation of the carrier frequency, above and below its unmodulated centre frequency, is given by the term $\pm K V_m$. This same peak variation is called the frequency deviation, f_d.

Therefore, $f_d = KV_m$, and substituting this into the previous equation for the instantaneous frequency of a frequency modulated carrier wave gives

$$f = f_c + f_d \sin \rho t \qquad (10.5)$$

We have seen above that the expression for the instantaneous carrier wave voltage is given by

$$v_c = V_c \sin \theta \qquad (10.6)$$

and that for the phase angle by

$$\theta = 2\pi \int f \, dt \qquad (10.7)$$

We can develop Equation 10.7 by combining it with Equation 10.5 and obtaining

$$\theta = \omega_c t - \eta \cos \rho t$$

Because f_d/f_m is the modulation index, η, we can say

$$\theta = 2\pi \int (f_c + f_d \sin \rho t) dt$$

$$= 2\pi \left(f_c t - \frac{f_d}{\rho} \cos \rho t \right)$$

$$= 2\pi f_c t - \frac{2\pi f_d}{2\pi f_m} \cos \rho t$$

where $2\pi f_m = \rho$

$$\theta = \omega_c t - \eta \cos \rho t$$

or

$$\theta = \theta - \eta \cos \rho t \qquad (10.8)$$

θ_c in radians is the carrier phase angle and $\eta \cos \rho t$ is the phase deviation in radians.

In a similar manner we can combine Equations 10.6 and 10.7 and obtain the following expression for the instantaneous voltage of a frequency modulated carrier wave:

$$v = V_c \sin \left(2\pi \int f \, dt \right)$$

Substituting for f from Equation 10.3 we get

$$v = V_c \sin\left[2\pi \int (f_c + f_d \sin \rho t) dt\right]$$

$$v = V_c \sin\left[2\pi \left(f_c t - \frac{f_d \cos \rho t}{\rho}\right)\right]$$

$$v = V_c \sin\left(2\pi f_c t - \frac{2\pi f_d}{\rho} \cos \rho t\right)$$

However, since $2\pi f_c = \omega_c$ and $\rho = 2\pi f_m$, where f_m is the modulating signal frequency, we can say

$$v = V_c \sin\left(\omega_c t - \frac{f_d}{f_m} \cos \rho t\right)$$

Since the modulating index $\eta = f_d/f_m$, we can rewrite the expression for the instantaneous voltage of a frequency modulated wave as:

$$v = V_c \sin(\omega_c t - \eta \cos \rho t) \tag{10.9}$$

10.7.3 The frequency spectrum of a frequency modulated wave

Since $\sin (A - B) = \sin A \cos B - \cos A \sin B$, we can expand Equation 10.9 as follows:

$$v = V_c \sin(\omega_c t - \eta \cos \rho t)$$

$$v = V_c \left[\sin \omega_c t \cos(\eta \cos \rho t) - \cos \omega_c t \sin(\eta \cos \rho t)\right]$$

Further trigonometrical manipulation, using what is known as Bessel's law, enables us to translate the two terms containing the modulating angular frequency, $n \sin \rho t$ and $n \cos \rho t$, into a multiplicity of higher frequency terms of $\sin \rho t$, $\cos \rho t$, $\sin 2\rho t$, $\cos 2\rho t$, $\sin 3\rho t$, $\cos 3\rho t$ and so on. In engineering terms, this means that a sinusoidal carrier wave of frequency f_c, if frequency modulated by another sinusoidal wave of frequency f_m, will produce a complex spectrum of waves. The frequency spectrum produced will consist of the original carrier frequency together with a number of side frequencies equally disposed about the central carrier frequency. These side frequencies are found to be $(f_c \pm f_m)$, $(f_c \pm 2f_m)$, $(f_c \pm 3f_m)$, etc., and are therefore spaced at whole multiples of f_m on each side of f_c. For example, a 1000 Hz modulating frequency produces a first pair of side frequencies spaced 1000 Hz above and below f_c; a second pair spaced at 2000 Hz around f_c and so on with the nth pair being at $(f_c \pm 1000n)$ Hz.

In theory, frequency modulation produces an unlimited number of side frequencies. But the amplitude of some of the side frequencies is insignificant such that they can be ignored. If we plot a graph of the various side frequency amplitudes against the frequency modulation index and, for comparative purposes, make the carrier frequency amplitude unity, we obtain Figure 10.19. From this graph we can make two observations:

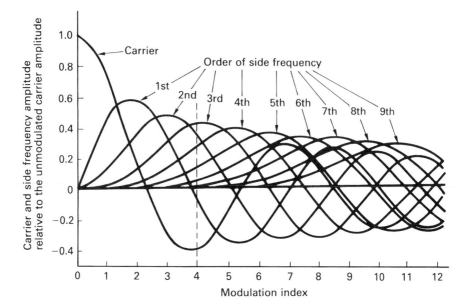

Figure 10.19 *Diagram to show how the carrier and side frequency amplitudes vary with changes in modulation index*

- at certain values of the modulation index the amplitude of the carrier wave, or one or more of the side frequencies, is zero; and
- when the modulation index is less than about 0.5, the number of side frequencies involved is only two and hence we can deduce that for small values of the modulation index the bandwidth requirement of the f.m. wave is virtually the same as that of the a.m. wave.

Figure 10.19 can be used to determine the frequency spectrum of a f.m. wave for a particular modulation index. Figure 10.20 is the frequency spectrum for a modulation index of 4. This has been obtained by plotting the individual amplitudes of the carrier wave and its side frequencies as given by their intersection with a vertical line drawn through the point where the modulation index is 4 in Figure 10.19.

If the data to be sent by a f.m. wave are not to be distorted, the carrier wave and all of its side frequencies should be transmitted. However, in the interest of limiting the bandwidth required for the transmission, the following empirical formula can be used to calculate a reduced bandwidth, B, which still produces satisfactory practical results:

$$B = 2(f_d + f_m) \text{ (Hz)}$$

where f_d is the frequency deviation, and f_m is the modulating frequency.

This equation can be verified by considering Figure 10.19 for a modulation index of 4, which we have seen produces the frequency spectrum shown in

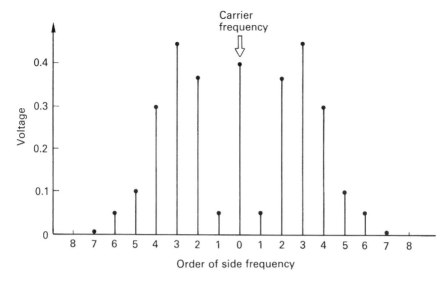

Figure 10.20 *Frequency modulation side frequency spectrum for a modulation index of 4. The voltage scale shows the proportion of the unmodulated carrier amplitude*

Figure 10.20. For example, a modulation index of 4 would be produced by $f_m = 15$ kHz together with $f_d = 60$ kHz. Using the formula the bandwidth required for satisfactory transmission is:

$$B = 2(15 + 60)\ \text{kHz} = 150\ \text{kHz}$$

We can see from Figure 10.20 that for a modulation index of 4 the highest order side frequencies of significant amplitude is the sixth pair. Therefore the bandwidth required to embrace the upper and lower sixth order side frequencies is $2 \times 5 \times 15$ kHz $= 150$ kHz. This matches the value for B obtained with the empirical formula.

If the modulating frequency is not a simple single sine wave, as we have considered thus far, but rather a complex wave, then the empirical formula still applies for the calculation of B provided the *highest* modulating frequency is used for f_m.

10.8 Modulation methods

There are many different circuit designs for the process of amplitude or frequency modulating a carrier wave but only one typical example of each will be considered here.

10.8.1 The square law amplitude modulator

A circuit for this function is shown in Figure 10.21. The transistor is required to have a non-linear, square law transfer characteristic. The instantaneous current flowing in the collector circuit is then of the form:

$$i_{mc} = I_o + a_1 v_{in} + a_2(v_{in})^2 + a_3(v_{in})^3 + \dots$$

If we replace v_{in} by $(V_c \sin \omega t + V_m \sin \rho t)$, being the instantaneous sum of the carrier and modulating waves, we can expand the expression using trigono-metric identities. The result is a series of terms some of which represent a d.c. component while others represent a range of component currents at different frequencies. One such set of terms is:

$$a_1 V_c \sin \omega t + a_2 V_c V_m [\sin(\omega - \rho)t + \sin(\omega + \rho)t]$$

and this represents the carrier wave and its upper and lower side frequencies. In Figure 10.21 the voltages represented by these terms are separated from the unwanted d.c. and higher frequency harmonic components by the LC collector load which is tuned to ω. The separation process is helped if v_{in} is kept small as this reduces the magnitude of the unwanted voltage components in proportion to the wanted ones.

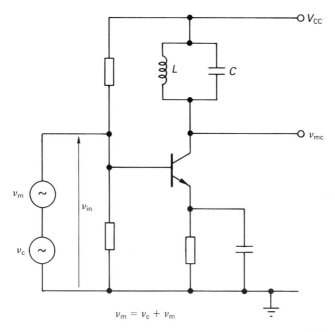

Figure 10.21 *Amplitude modulation using a 'square law' modulator circuit:*
$$V_{in} = V_c + V_m$$

10.8.2 Frequency variation using a reactance diode

Figure 10.22 shows the principle of using the inherent voltage sensitive capacitive effect of a diode to change the running frequency of an oscillator. The basic oscillator frequency is determined by the tuned circuit comprising L and C. This is shunted by the variable capacitance formed by the *reactance diode*, C_d. The modulating voltage, v_m, is applied to the cathode of C_d, thus causing its capacitive effect to vary at the modulating voltage frequency and by an amount determined by its amplitude. The oscillator frequency is therefore frequency modulated and produces the modulated carrier voltage output, v_{mc}.

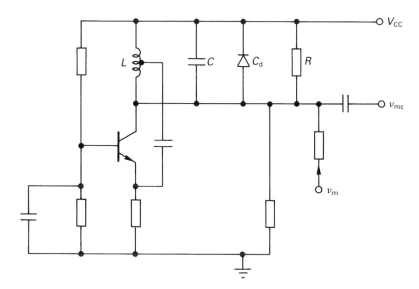

Figure 10.22 *A method of frequency modulating an oscillator using a 'reactance diode'. C_d acts as a voltage controlled capacitor across the oscillator tuned circuit LC operating at the nominal carrier frequency. v_m is the modulating voltage, v_{mc} is the modulated carrier output*

10.9 The principles of demodulation

To conclude this chapter we shall take only a brief simplistic look at the principles involved in *demodulating* or extracting the intelligence from the two types of modulated carrier voltage waveforms we have considered earlier.

10.9.1 Amplitude modulation

The principle of amplitude demodulation is illustrated in Figure 10.23. The idea is to remove the carrier frequency, f_c, from the modulated carrier, v_{mc},

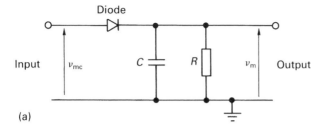

(a)

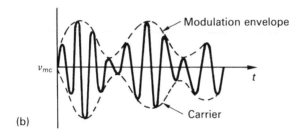

(b)

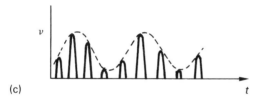

(c)

(d)

Figure 10.23 *The principle of amplitude demodulation. (a) The simple detector circuit. (b) The modulated carrier input waveform. (c) The effect of the diode. (d) The output waveform.*

leaving only the modulation voltage, v_m, which contains the information being transmitted.

The simplest demodulation circuit comprises a diode which first removes the negative half cycles of the modulated carrier. This is followed by the capacitor–resistor combination, which acts as a filter or smoothing circuit ($CR \gg 1/f_c$) effectively to produce a unidirectional voltage, v_m, varying at the envelope or modulation frequency.

10.9.2 Frequency modulation

A simple form of frequency demodulator could consist simply of a transistor having an *LC* tuned circuit as its collector load. At the resonant frequency of the tuned circuit its impedance would be purely resistive and equal to L/CR ohms. A signal of resonant frequency applied to the base of the transistor would thus produce the maximum r.m.s. output voltage across the tuned circuit. Applied signal frequencies higher or lower than that of resonance would undergo less gain and result in a reduced r.m.s. voltage output. If the applied signal were a frequency modulated wave, having an unmodulated centre frequency, f_c, less than the tuned circuit resonant frequency, then variations in frequency about f_c would cause sympathetic variations in output voltage amplitude. In other words, the frequency modulation would have been converted into amplitude modulation. This being the case, the extraction of the data signal is then simply a matter of detecting the amplitude modulation as outlined in the previous section. Figure 10.24 shows the tuned circuit response curve and how a signal off-tuned to either X or Y from the resonant frequency f_o will produce amplitude variations. In order that the translation from frequency to amplitude variations is as accurate as possible, only the linear part of the response curve should be used. This implies keeping the input signal frequency swing to a minimum.

An improvement on the single tuned circuit frequency discriminator or demodulator can be obtained by using a pair of tuned circuits. Such a circuit arrangement is shown in Figure 10.25(a). The two *LC* circuits are stagger-tuned around the carrier centre frequency and are connected to produce

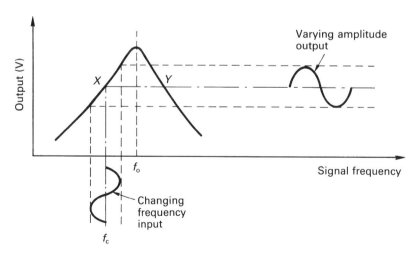

Figure 10.24 *A tuned circuit can be used to transform frequency variations into amplitude changes.* f_c, *Carrier centre frequency;* f_o, *tuned circuit resonance frequency*

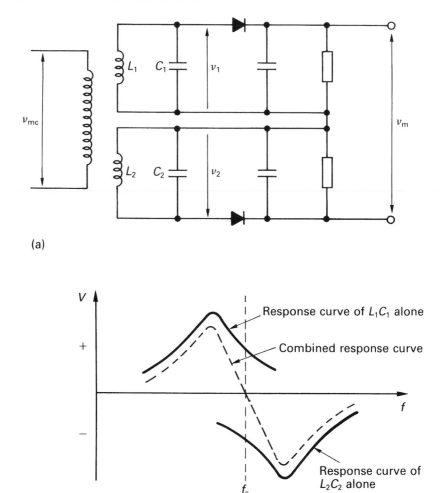

(a)

(b)

Figure 10.25 *Frequency demodulation. L_1C_1 is tuned to resonate below the carrier frequency and L_2C_2 above the carrier frequency, f_c. (a) A f.m. demodulator circuit. (b) The f.m. demodulator response curve*

voltages which are in phase opposition. At the centre frequency, f_c, the two tuned circuits produce equal and opposite voltages ($v_1 = v_2$) and the output voltage, v_m, is zero. For input frequencies of v_{mc} above or below f_c, v_1 and v_2 are unequal and add algebraically together to produce a more linear combined response as is shown by the dotted line in Figure 10.25(b). The frequency variations of the modulated carrier input voltage, v_{mc}, around f_c, thus result in the output data signal amplitude, v_m, varying at the same rate.

Exercises

10.1 A 40 channel, pulse code modulated data link uses an 8-bit word length to describe the analogue amplitudes of the data signals it is carrying. If the sampling rate is 6000 samples per second, calculate the link transmission bit rate.

10.2 A 10 kHz sinusoidal carrier wave is amplitude modulated to a depth of 50% by a modulating signal comprising: (a) a sinusoidal waveform at a frequency of 1 kHz, and (b) a square waveform with a period of 1 ms. For each case, (a) and (b) above, draw dimensioned sketches of the modulated carrier waveform you would expect if these were shown on the screen of an oscilloscope.

10.3 Suppose a carrier wave, given by $v_c = 1.0 \sin 8000\pi t$, is amplitude modulated by a signal given by $v_m = 0.5 \sin 600\pi t - 0.125 \cos 1200\pi t$. Show that the resulting modulated carrier wave is given by the equation $v_{mc} = \sin 8000\pi t + 0.25 \cos 7400\pi t - 0.25 \cos 8600\pi t - 0.0625 \sin 9200\pi t - 0.0625 \sin 6800\pi t$.

10.4 (a) Compare the bandwidth used by a double side band a.m. wave with that used by a f.m. wave having a modulation index of 4, a modulating frequency of 2 kHz and a carrier frequency of 1.5 MHz.

(b) Estimate the new bandwidth of the f.m. signal if the modulation index is doubled.

10.5 A carrier wave is frequency modulated by a 3 kHz sinusoidal signal such that a modulation index of 0.5 is obtained. Estimate:

(a) the bandwidth required for satisfactory transmission; and

(b) the change in bandwidth if the modulation index is increased to 5.

10.6 A 400 Hz sinusoidal carrier wave is frequency modulated by a signal of square waveform and period 20 ms. It is arranged that this produces maximum carrier frequency deviations of ±50 Hz.

(a) Sketch the waveform of the modulated carrier, taking care to indicate the correct number of cycles corresponding to the instantaneous signal conditions.

(b) Suppose that the shape of the modulating signal is changed from square to sinusoidal but with the same amplitude and frequency being retained.

Write down a mathematical expression for the instantaneous frequency of this new modulated carrier and calculate the modulation index.

10.7 A carrier wave, $100 \sin 8\pi \times 10^6 \, t$ volts, is frequency modulated by a signal of amplitude 2 volts and frequency 1 kHz.

(a) If the modulating signal amplitude produces a frequency deviation of 80 kHz, write down a mathematical expression for the instantaneous voltage of the modulated carrier wave.

(b) Use the expression from (a) to derive a further expression for the instantaneous frequency of the modulated carrier.

(c) Calculate the maximum and minimum frequencies of the modulated carrier wave.

11
Analogue and digital conversions

11.1 Introduction

There is an ever increasing use of digital computers in modern process control systems and the transmission of information is increasingly undertaken using digital communication systems. However, the analogue 'ends' of the typical digital control or communications system have not been outmoded; they never will be so long as people need to interface with them. For this reason alone, there is an important requirement for the rapid and accurate translation between the analogue and digital formats.

Figure 11.1 is a block diagram of a typical analogue process which employs a digital microcomputer. The sensing transducer in the analogue process may be a thermocouple responding to the temperature of a tank of heated liquid. The temperature of the tank could be monitored by the digital microcomputer. It could also be required to compare the reading from the thermocouple with a pre-set value it already holds. The result of the comparison may require the microcomputer to send control signals to an analogue fuel supply valve back in the analogue process. Such a situation requires both an analogue-to-digital conversion (ADC) and the reverse digital-to-analogue (DAC) function.

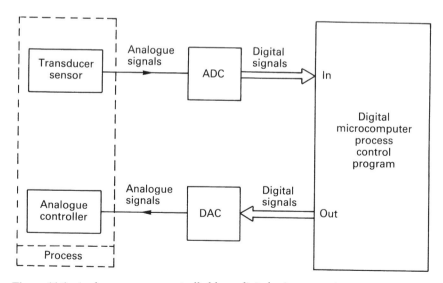

Figure 11.1 *Analogue process controlled by a digital microcomputer*

11.2 General conversion principles and terminology

11.2.1 Sampling the analogue input

Figure 11.2 shows the principle involved in sampling an analogue voltage to form a series of digital pulses. The analogue waveform to be sampled is shown in (a), the times at which (a) is sampled in (b) and the train of sample digital pulses so produced in (c). The height of the sample pulses represents the instantaneous amplitude of the analogue waveform at the instant of being sampled. A line joining the tops of the sampled pulses traces the form of the original analogue waveform. The greater the number of samples in a given period of time, the more accurately will the sampled pulses represent the original waveform.

Clearly, the fewer times we need to sample the analogue waveform, the less the overall time needed to produce a representative digital sample. However, for satisfactory conversion results, the analogue signal must be sampled at a rate which is at least twice that of the highest frequency in the analogue waveform. This is known as the *Nyquist sampling rate*. Nyquist showed that if at least two regularly timed samples can be taken of each cycle of the highest input analogue frequency then the reconstructed waveform will contain the same information as the original.

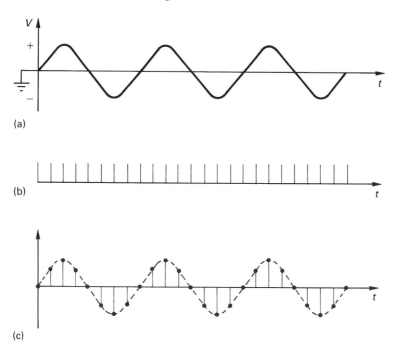

Figure 11.2 *Amplitude time sampling. (a) Waveform to be sampled. (b) Equal interval sampling times. (c) Sampled waveform amplitudes*

11.2.2 Aliasing

Figure 11.3 shows the effect known as *aliasing*. This occurs when the sampling rate is lower than the Nyquist rate. Figure 11.3(a) shows the case where a triangular analogue waveform of 8 kHz is being sampled at a rate of only 10 kHz rather than at least twice 8 kHz, namely 16 kHz. The digital sample obtained, when reconstituted, may well produce another triangular wave but its frequency is lower than that of the original. Even when the sampling rate is increased above the Nyquist rate, there is still a multiplicity of frequencies which will produce the same digitised sample. Figure 11.3(b) shows this effect with two sinusoidal waveforms. In order to prevent interference and confusion with any unwanted higher frequency signals, anti-aliasing filters are used in the input analogue signal path.

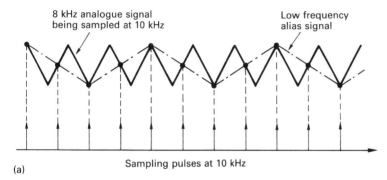

(a)

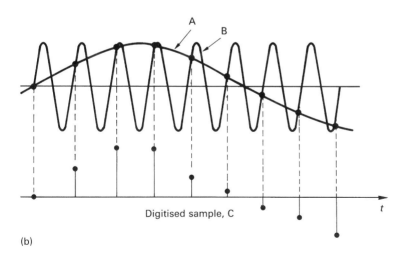

(b)

Figure 11.3 *Aliasing. (a) If the sampling rate is too low, a false (alias) impression of the sampled waveform is obtained. (b) Both waveforms A and B can be represented by the same digitised sample, C*

11.2.3 The sample and hold circuit

A typical circuit is shown in Figure 11.4. The analogue voltage sample is stored in the low leakage hold capacitor, C. The metal-oxide semiconductor field-effect transistor (MOSFET) is no more than a pulse controlled, low resistance electronic switch which connects the capacitor to analogue input for a short period of time – the *sampling time* or the *aperture time*. The purpose of the high input impedance buffer, A1, is to provide the necessary capacitor charging current, so preventing the sampling circuit from loading the analogue voltage source. During the sampling time the hold capacitor charges to the voltage level of the sampled analogue waveform. The time taken for the capacitor to become fully charged by the sampled voltage level is known as the *acquisition time*. The purpose of the second buffer amplifier, A2, is to reduce any current drain from the hold capacitor during its 'hold' mode after the sample has been collected. Any current drain from the hold capacitor after the sample voltage has been stored is undesirable; it causes the sample analogue voltage to 'droop' during its conversion to digital form. In practice, there is always a fall in the hold voltage across the capacitor; besides the inevitable small current taken by the buffer amplifier, there are additional contributions by the leakage currents of the MOSFET, switch and the capacitor dielectric. A further practical point is that the selection of the sampling or aperture time needs care. If it is too short, the hold capacitor will have insufficient time to charge fully to the sampled voltage.

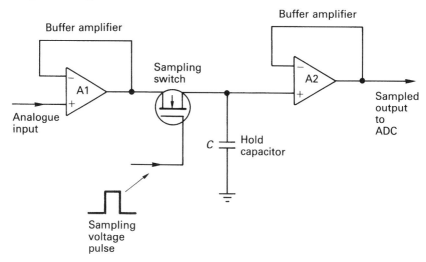

Figure 11.4 *Typical sample and hold arrangement*

11.2.4 Conversion

The sample and hold circuit presents its constant voltage sample to the analogue-to-digital converter (ADC) for the final stage of the conversion process. The usual way of converting the different analogue samples is to divide the total analogue range into a number of voltage levels. Each level is given a digital code which is also allocated to any analogue voltage sample which happens to correspond closest to that level.

For example, suppose we decided to use a 3-bit digital code to represent the full range of the analogue voltages which we wished to convert. There is a total of eight unique levels which can be represented by '0' and '1' in groups of three digits; that is, from 000 to 111. Also suppose the full analogue voltage range to be 0 V to 3.5 mV. Table 11.1 shows the full situation.

Table 11.1 *Analogue conversion to 3-bit digital*

Level No.	*Sample* (mV)	*Binary code*
1	0	0 0 0
2	0.5	0 0 1
3	1.0	0 1 0
4	1.5	0 1 1
5	2.0	1 0 0
6	2.5	1 0 1
7	3.0	1 1 0
8	3.5	1 1 1

The range or spread of each voltage level is 0.5 mV. This method of allocating all sample analogue voltages the binary code of their nearest standard level is called *quantising* and using a 3-bit converter there are eight levels having, in this case, a *quantum* of 0.5 mV. The *resolution* of a measuring system is defined as the smallest input which the system will detect. Thus, the quantum figure of 0.5 mV is also the smallest input which this 3-bit ADC can accurately represent.

Because the values of analogue voltage to be converted will generally not lie on an exact quantum level, there will be an uncertainty as to the correctness of the least significant bit. Referring to Table 11.1, the digital code 010 represents any analogue voltage from 0.75 mV to just less than 1.25 mV; that is, 1.0 mV $\pm$ 0.25 mV. Therefore, the ambiguity or error is seen to be half a quantum level and it is known as the *quantising error* or the ± 1 error.

Unfortunately, there is no way of removing the uncertainty of the least significant bit. But the amount of uncertainty can be reduced by the use of more than three bits in the digital conversion code. In general, the resolution or quantum level for a binary coded conversion is given by the expression:

Resolution = Analogue voltage range$/(2^N - 1)$

where N is the number of bits used.

For example, by using 4 bits, the ± 0.25 mV error of the 3-bit system in Table 11.1 is reduced to

$$3.5/2^4 - 1 = 3.5/15 = 0.233 \text{ or } \pm 0.117 \text{ mV}$$

For an 8-bit system the error becomes ± 0.00686 mV while a 12-bit arrangement reduces the error to ± 0.000427 mV.

Figure 11.5 shows the ideal classic 'staircase' transfer characteristic for a 4-bit ADC. The lowest analogue voltage level is quantised as 0000 and there are 15 subsequent equal steps, each incrementing by 0001, to reach the 16th quantisation level having the digital number 1111.

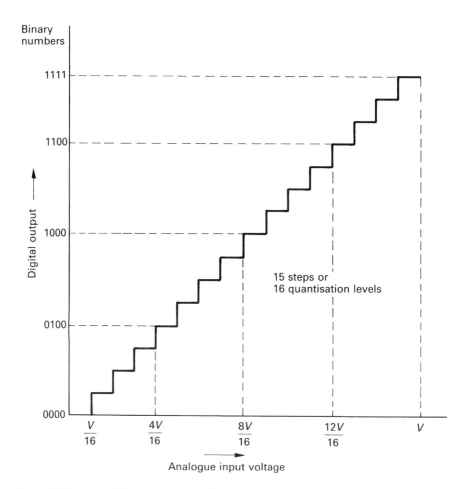

Figure 11.5 *4-bit ADC 'staircase' transfer characteristic*

11.2.5 Quantisation noise

It is interesting to note that the process of converting from analogue to digital representation is not completely reversible. Figure 11.6 illustrates this point. The continuously variable analogue waveform in (a) is translated into discrete samples in (b). However, if an attempt is made to reconstitute the original waveform in (a) from the samples in (b), the result is as shown in (c). The smoothness of the original analogue waveform has given way to a jerky, 'castellated' shape. The sharp steps up and down constitute unwanted high frequency interference commonly known as *quantisation noise*.

(a)

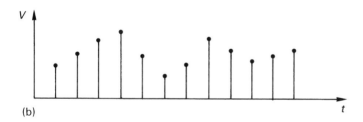

(b)

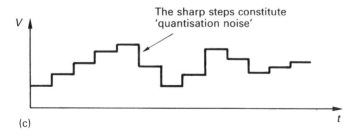

(c)

Figure 11.6 *The sampling process is not completely reversible. (a) Original analogue voltage waveform. (b) Discrete samples of (a). (c) Analogue waveform formed from (b)*

11.3 Digital-to-analogue conversion

We shall see later, a little surprisingly so, that some of the more popular ADC processes themselves use a DAC device. For this reason alone it is important that before ADC is considered any further we have a closer look at the principles of DAC.

Of the two conversion processes, the understanding and hardware implementation of DAC is simpler than that for ADC. Of the various DAC methods available, there are only two of major importance and both these use an operational amplifier in a current summing role. Before these two methods of converting digital signals to analogue waveforms are explained, a brief comment on resolution as applied to this process would be useful.

The input to the DAC is a number of bits, in a certain pattern, which constitute a *digital word*. The greater the number of bits the better the resolution and the more finely can the analogue output represent the digital input. The resolution of the DAC system is decided by the effect of changing the least significant bit (LSB) of the input digital word. A 4-bit word input will produce 16 different levels of analogue output for a given applied reference voltage supplied to the DAC; an 8-bit input will have 256 different output levels for the same DAC reference voltage. The degree of conversion uncertainty is therefore decided by the number of bits in the input because the analogue output settles to within $\pm$ half of the LSB.

We shall see later that the DAC is an extremely fast operator. Its speed of conversion is limited by what is known as its *settling time*. The shortest conversion time occurs when the digital word changes by a single LSB. In the case of the 8-bit converter, this means the analogue output has to change by only 1/255 of the reference voltage. The longest settling time is required when all the digits of the input word change, say from 01111111 to 10000000. If all eight digits change, a typical settling time can be as long as a microsecond.

11.3.1 The weighted resistor method

The circuit for a 4-bit DAC is shown in Figure 11.7. An operational amplifier is supplied at its summing point with four currents the amplitudes of which have a binary relationship. The currents are sourced by a stable reference voltage, V_R, and a bank of four electronic switches. The switches are controlled by digital pulses which are applied as a parallel 4-bit digital word. The digital word represents the analogue voltage to be retrieved from the output of the DAC. The bank of switches connects either an earth or V_R to the appropriate operational amplifier input lines. Currents I_1 to I_4 flow through resistors R_1 to R_4 towards the summing point marked V_X. The resistors are sized such that current $I_1 = 2I_2 = 4I_3 = 8I_4$. Since no current flows into the terminals of an ideal operational amplifier, the sum of I_1 to I_4 flows through R_f. Because V_X is a virtual earth, the volt drop caused by the current flowing through R_f is in fact e_o, the required analogue conversion.

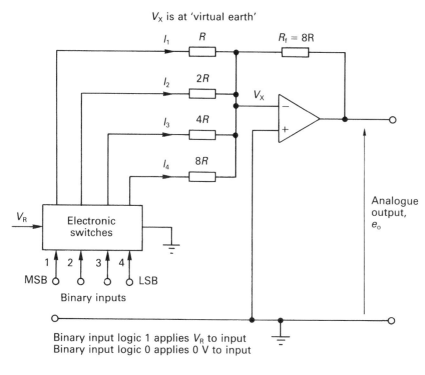

V_X is at 'virtual earth'

Figure 11.7 *Weighted resistor 4-bit DAC*

Binary input word	e_o
1 1 1 1	15 V_R
1 0 1 0	10 V_R
0 0 1 1	3 V_R
. . . .	.
. . . .	.
0 0 0 0	0

Suppose the binary inputs from the most significant bit (MSB) to the LSB are 1010 respectively. This causes the electronic switch to apply V_R to input lines 1 and 3 and earth to lines 2 and 4. Because V_X is a virtual earth we can say:

$$I_1 = V_R/R$$

$$I_3 = V_R/4R$$

$$I_2 = I_4 = 0$$

Now

$I_f = I_1 + I_2 + I_3 + I_4$

$= V_R/R + 0 + V_R/4R + 0$

$= 5V_R/4R$

Suppose that the value of R_f has been chosen to be $8R$ such that

$e_o = I_f R_f$

$= (5V_R/4R)8R$

$= \mathbf{10}V_R$

Similarly, if the digital input were 0011, then

$I_f = 0 + 0 + V_R/4R + V_R/8R$

$= 3V_R/8R$

This gives a value for $e_o = I_f R_f$:

$e_o = (3V_R/8R)8R$

$= \mathbf{3}V_R$

A disadvantage of using this circuit is the range of different values of resistor required. This is not so apparent with the 4-bit example used above but if this design were to be extended to an 8-bit or 12-bit DAC it could become a problem to manufacture. The modern way of obtaining a DAC is to buy a commercially available integrated circuit design. These almost invariably use the R–$2R$ method outlined below because it requires only two different resistor values.

11.3.2 The 4-bit R–$2R$ ladder DAC

The circuit for this device is shown in Figure 11.8(a). Again, the principle is to feed the summing terminal of an inverting operational amplifier with four currents having a binary relationship. The currents are sourced through a solid state bank of switches, S1 to S4. The switch is controlled by the 4-bit digital word which represents the voltage requiring conversion to analogue form. The switches connect to earth if the bit input is a logic 0, but to the operational amplifier summing point, V_x, if it is a logic 1. (Note that V_X is a virtual earth which is what makes all this possible.) The result is a binary pattern of input currents which flow through the resistor ladder network to the operational amplifier summing point. Since no current enters the input terminal of the ideal operational amplifier, the sum of the binary related currents effectively flows through R_f, causing the voltage drop which is the output voltage, e_o.

The cleverness of this R–$2R$ ladder network design is its ability to accept increased lengths of the input digital word by the addition of an R–$2R$ section for each additional input digit. The way this is possible is illustrated by considering the simple network shown in Figure 11.8(b). The resistive path

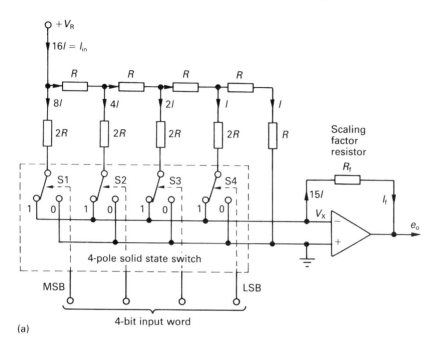

(a)

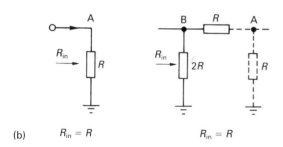

(b) $R_{in} = R$ $R_{in} = R$

Figure 11.8 *The 4-bit R–2R ladder DAC. (a) The basic circuit:* $e_o = I_f R_f$. *(b) The operation of the R–2R ladder. The addition of R–2R sections does not change the circuit input resistance to earth*

from point A to earth is R. The addition of the R–$2R$ section to earth does not alter the resistance of the changed path to earth; from the new viewpoint, B, it is still R. The addition of further R–$2R$ sections allows the simple addition of further digits to the input word. Each new R–$2R$ section merely causes the equal (binary) division of any current flowing into its junction. Reference to Figure 11.8(a) shows the binary division of the currents in the various limbs of the R–$2R$ ladder network.

With the binary input being 1111, as is depicted by Figure 11.8(a), and assuming the input current from the reference voltage source to be $16I$, for the sake of argument, the total current flowing through the scaling resistor, R_f, is $15I$. This produces an output voltage

$$e_o = 15IR_f$$

If the binary input word were 1010 the summing current would be $10I$; for 1000, $8I$; for 0101, $5I$ and so on.

Since the R–$2R$ ladder network always presents V_R with a resistance R to earth, the input current, I_{in}, from the reference source will always be

$$I_{in} = V_R/R$$

For the 4-bit DAC shown, $I_{in} = 16I$ in which case

$$16I = V_R/R \text{ or } I = V_R/16R$$

Now, if for example the digital input word were 1000, then $e_o = 8IR_f$ could be written as

$$e_o = 8R_f(V_R/16R)$$

If we were now to make $R_f = R$ and $V_R = 16$ V, then e_o would be a convenient 8 V, so matching the denary equivalent of the binary input.

Thus, by selecting suitable values for V_R and R_f, the R–$2R$ DAC can be scaled to give difference ranges of analogue output voltages. However, remember that the operational amplifier has its output voltage swing curtailed by its d.c. supply voltage. This being the case it is possible for V_R to be set too high for the operational amplifier to follow. For example, if V_R were 16 V and the binary input word 1111, the R–$2R$ ladder network would attempt to force e_o to 15 V. But, if the operational amplifier supply rail were only 12 V, e_o would be limited also to 12 V; it would be incapable of following digital word inputs greater than 1100.

11.4 Analogue-to-digital conversion

11.4.1 The 'flash' ADC

Figure 11.9 shows a typical circuit arrangement for an analogue to 2-bit conversion. The analogue voltage to be converted is compared with a standard or reference voltage until as near a match as possible is obtained. The analogue voltage to be converted is fed in parallel to the inverting terminal of each of a bank of comparators. The non-inverting terminals are each connected to a different point on a potential divider chain formed by a series of like value resistors, R, across a reference voltage, V, and earth. Figure 10.9 shows a 4-step voltage divider chain and hence comparator A will switch its output voltage from logic 0 to logic 1 when the analogue voltage input just exceeds $V/4$. The other two comparator outputs remain at logic 0 and the digital input to the encoder circuit block from the comparators A, B and C is respectively 1, 0 and 0. This digital input word changes to 110 and finally 111 as the analogue input voltage increases progressively. The table with Figure 11.9 illustrates the four

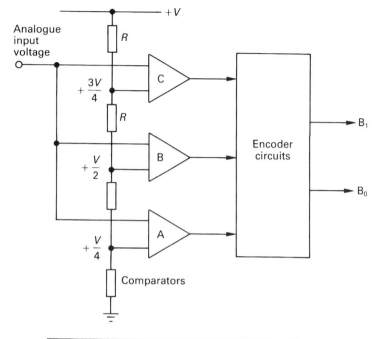

Analogue input voltage	Comparator outputs			Binary output	
	A	B	C	B_0	B_1
0 to $\frac{V}{4}$	0	0	0	0	0
$\frac{V}{4}$ to $\frac{V}{2}$	1	0	0	0	1
$\frac{V}{2}$ to $\frac{3V}{4}$	1	1	0	1	0
$\frac{3V}{4}$ to V	1	1	1	1	1

Figure 11.9 *Flash ADC*

ranges of analogue input and the four different 2-bit outputs they generate from the encoder circuits.

This type of ADC is called a 'flash' converter because it is the fastest available; the conversion delay can be less than 20 ns. It is readily available in integrated circuit form with 4-bit to 8-bit outputs. A 2-bit converter is shown in Figure 11.9 as requiring three comparators. An 8-bit device, with 256 levels of output to define, would require 255 comparators. It is this complexity which makes the manufacturing costs and the size of the device too large for higher bit outputs.

11.4.2 The successive approximation ADC

Figure 11.10 shows the block diagram of a 4-bit successive approximation ADC. The principle here is to feed a succession of trial digital words to a DAC the analogue output of which is compared with the analogue voltage requiring conversion. When the two are equal, the digital word input to the DAC is the required digital conversion. The analogue voltage sample, V_1, is compared in the comparator with the output, V_2, from the DAC. The comparator output is low until V_1 just exceeds V_2 when the comparator output switches to its high state.

Before a conversion process is started, the 4-bit programmable register is reset to zero, that is the four input bits to the DAC are all at logic 0. On receipt of the 'start conversion' signal from the clock, the programmed register next makes a trial input to the DAC by changing bit b_4, the MSB, to logic 1. The new digital input of 1000 to the DAC changes its output to the comparator

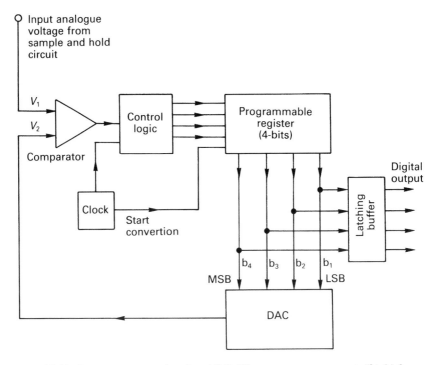

Figure 11.10 *Successive approximation ADC. The comparator output is 'high' for* $V_1 > V_2$ *and 'low' for* $V_1 < V_2$

from zero to just half of the DAC reference voltage. This voltage, V_2, is compared with V_1 and the resulting comparator output of either a logic 0 or a logic 1 tells the control logic whether the first digital 'guess' of 1000 is too high or too low. If too low, for example, b_4 is left set at logic 1, as illustrated by Figure 11.11, otherwise it is reset to logic 0. The next clock pulse causes the programmable register to change b_3 to a logic 1 and the trial input to the DAC becomes 1100. If the resulting analogue output from the DAC is now greater than V_1, b_3 is reset to logic 0 and b_2 set to logic 1 to produce the third digital trial of 1010. Figure 11.11 shows this to be too low, and so on the fourth clock pulse b_1 is set to logic 1, giving a final trial of 1011. In our example this is exactly correct at 5.5 V, but in practice, because of the rather coarse resolution of the 4-bit DAC, 1011 would also have been the final ADC conversion for an analogue input lying anywhere between 5 V and 6 V.

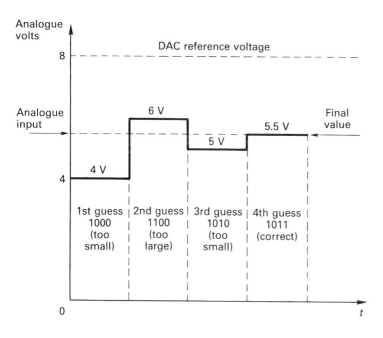

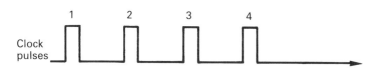

Figure 11.11 *4-bit successive approximation action. Conversion from the analogue voltage input (5.5 V) to the 4-bit digital output (1011) is completed in only 4 clock cycles, but the resolution is only $\frac{1}{2^4} = \frac{1}{16}$ of the full input range*

The successive approximation method conversion process is not so fast as the high performance flash converter but neither is it so complex. In integrated circuit form, for resolutions up to 12 bits, it has conversion times less than 50 μs and its accuracy is acceptable for most practical purposes.

11.4.3 The counter ramp ADC

Figure 11.12 shows the block diagram arrangement for this method which also uses a DAC. The analogue voltage, V_1, requiring conversion is fed to input terminal A of the comparator while terminal B receives an increasing stepped ramp voltage V_2, from the counter-fed DAC. Until V_2 exceeds V_1, the comparator output is at logic 1 and the AND gate is enabled, allowing the positive clock pulses to be passed to the binary counter. When V_2 exceeds V_1, the comparator feeds a logic 0 to the AND gate, causing the binary counter to stop. The DAC then ceases to ramp up and the analogue voltage, V_1, is represented by the digital counter setting which is preserved by the latched buffer. This situation remains until the control logic signals another 'start

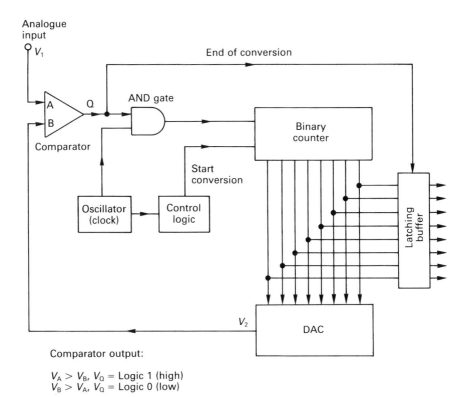

Comparator output:

$V_A > V_B$, V_Q = Logic 1 (high)
$V_B > V_A$, V_Q = Logic 0 (low)

Figure 11.12 *Counter ramp ADC*

conversion' causing the procedure to repeat. Figure 11.13 gives a graphical representation of the above conversion action.

11.4.4 The single slope ramp ADC

The principle behind this method is to use a capacitor, rather than a counter and DAC, to produce the ramp voltage against which the unknown input analogue voltage, V_{in}, is compared. Figure 11.14 shows a circuit

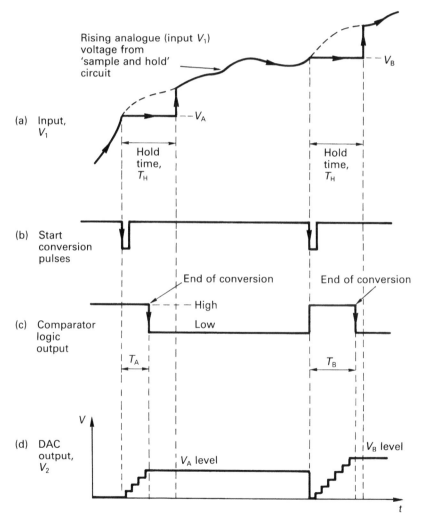

Figure 11.13 *Waveforms to explain the circuit shown in Figure 11.12. The acquisition times,* T_A *and* T_B, *must always be shorter than the hold time,* T_H

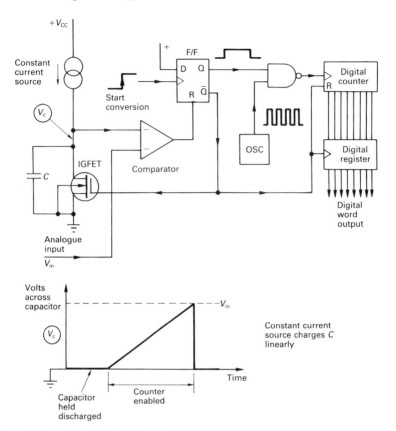

Figure 11.14 *Single slope ADC*

arrangement. *C* is the capacitor which is initially held in a discharged condition by the insulated-gate field-effect transistor (IGFET) which is conducting to earth because of the high potential on its gate terminal obtained from the D-type flip-flop lower output terminal. This same high output zeros the digital counter and associated register. V_c is at earth, and with $+V_{in}$, the comparator output is low.

With the onset of the leading edge of the 'start conversion' pulse, the flip-flop is enabled and the positive potential on its D input is transferred to and enables the following NAND gate, so passing a train of oscillator pulses to the counter. The IGFET in the meantime has been turned off by the now low NOT output from the flip-flop fed to its gate. This releases the capacitor from its earth clamp, allowing it to start to charge positively at a steady rate. The constant current charging source produces a linear increase of V_c which, when it just exceeds V_{in}, causes the comparator output to switch high and disable the flip-flop. This stops the counter and its reading at that time is the digital conversion of V_{in}.

The single slope integration method is still used in applications where a good resolution is more important than precise accuracy. The determination of the different heights of a series of pulses is a typically suitable application. It is also much used for the conversion of time to amplitude.

A disadvantage of the single slope method is its reliance for accuracy on the stability and accuracy of the capacitor and comparator. For this reason, where high accuracy is required, the 'dual slope' method is preferred.

11.4.5 The dual slope ADC

This is one of the most widely used ADCs. While it still uses a capacitor and comparator, the inadequacies of these components are effectively cancelled out. This is achieved by using them to time both the charging of the capacitor and its subsequent discharging. A block diagram of the dual slope arrangement is shown in Figure 11.15. The capacitor, C, is charged for a fixed time, T_1, by the analogue voltage to be converted, V_x. After T_1 the capacitor charge has reached V_1 and the capacitor input is automatically switched to the reference voltage, V_R. The capacitor's target charging voltage is now reversed from V_x to V_R. This causes the comparator input voltage to run down to zero over a period of time set by V_1. The conversion cycle is shown in Figure 11.16. The operation is justified mathematically as follows.

The integrator is driven for a fixed time T_1 to produce an input to the comparator of

$$V_1 = \frac{1}{RC} \int_0^{T_1} V_X \, dt$$

Assuming that V_x is constant over the conversion period this becomes

$$V_1 = \frac{1}{RC} T_1 V_X$$

After the charging period T_1, the comparator input is switched from V_1 to V_2 given by

$$V_2 = V_1 - \frac{1}{RC} \int_0^t V_R \, dt$$

and because V_R is a constant we can write

$$V_2 = \frac{1}{RC} T_1 V_X - \frac{1}{RC} t V_R$$

The digital counter in Figure 11.15 is started at the onset of T_1 and continues to count time until the comparator output changes, indicating that $V_2 = 0$. The counter now indicates the time t_x, and V_x will be given by

$$V_x = \frac{t_x}{T_1} V_R$$

Thus, t_x and V_x can be seen to have a linear relationship and t_x is independent of C and R in particular and the integrator characteristics in

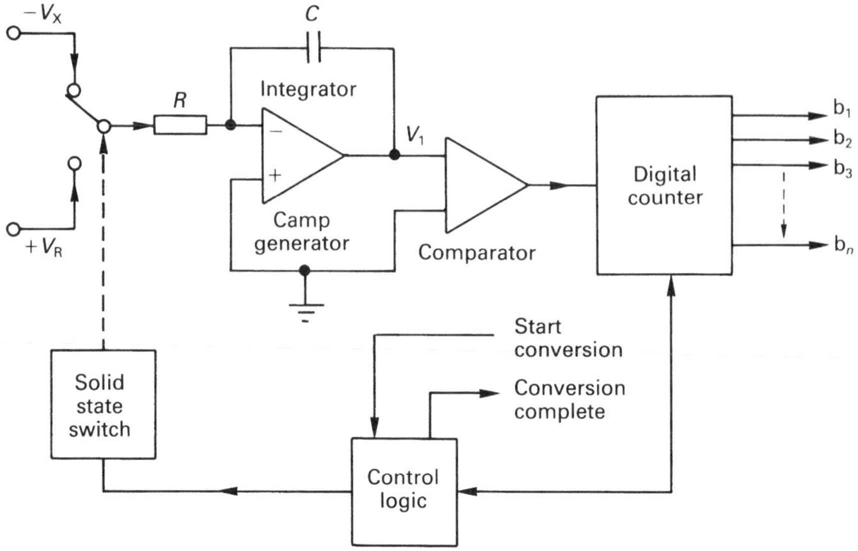

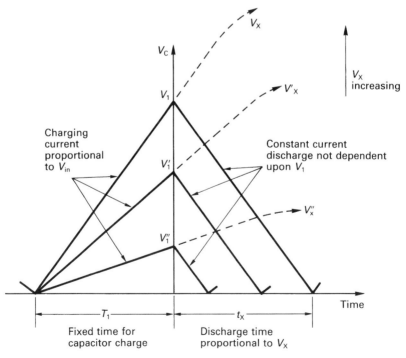

Figure 11.15 *Dual slope ADC*

Figure 11.16 *Dual slope conversion cycles for three different values of* V_X

general. The dual slope ADC system uses a *start* pulse and a *complete* pulse for each conversion.

Dual slope conversion is used extensively in precision digital multimeters and in integrated circuit modules having more than 10-bit resolutions. It gives good accuracy and high stability at an economic cost; it is however somewhat slow, with conversion times measured in tens of microseconds.

11.4.6 The delta–sigma charge-balance ADC

This is another method of compensating for lack of component stability. The principle involved here is the cancellation of the average analogue current input by a timed balancing current. The block diagram of this system is shown in Figure 11.17.

The analogue input voltage provides a charging current for the capacitor of an integrator the output of which is compared with earth. Depending upon the comparator output, current pulses of fixed time duration (Current × Time = Charge) are switched into the summing junction or to earth at each clock pulse. The net effect of this action is to maintain a zero current flow (i.e. a balanced charge flow) into the summing junction. The counter arrangement records the

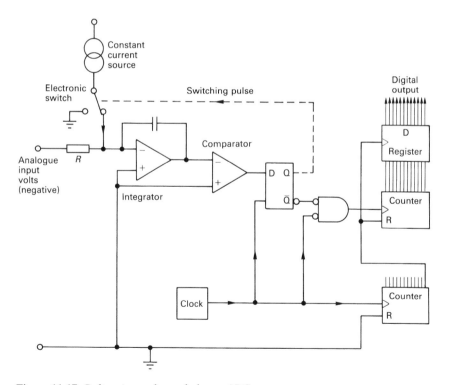

Figure 11.17 *Delta–sigma charge-balance ADC*

number, n, of fixed duration current pulses which are switched into the summing junction for a set number of clock pulses, N. The number of counts, n, is proportional to the average analogue input current during the period of N clock pulses. In other words, n is the digital conversion of the analogue input.

Exercises

11.1 A 7-bit DAC has the input word 1100111_2 and an 8 V reference.
(a) Find the analogue output voltage produced.
(b) Specify the conversion resolution voltage.

11.2 An 8-bit ADC has a reference voltage of 10 V and has an analogue input voltage of 6.875 V.
(a) Determine the digital output word.
(b) State the percentage resolution.

11.3 A 6-bit DAC is required to produce a 9 V output when its six input bits are all at logic 1. State the necessary reference voltage for this DAC.

11.4 The weighted 4-bit DAC shown in Figure 11.7 has a reference voltage of 1 V. The following table shows the input digital word and the corresponding analogue output.

Input	Output (V)	Input	Output (V)
0000	0	1000	8
0001	1	1001	9
0010	0	1010	8
0011	1	1011	9
0100	4	1100	12
0101	5	1101	13
0110	4	1110	12
0111	5	1111	13

(a) State which conversions are incorrect.
(b) Assuming the electronic switch is working correctly, state the most likely cause of the DAC functioning incorrectly.

11.5 The 4-bit R–$2R$ ladder DAC shown in Figure 11.8 has a voltage reference of 10 V, $R = 20$ kΩ, $R_f = 50$ kΩ and the operational amplifier output is bounded by its power supply to ± 18 V.
(a) If the input digital word is 1001, calculate the analogue voltage output.
(b) Determine the maximum digital input word that this DAC can truly convert.

11.6 The dual slope ADC as shown in Figure 11.15 has $R = 1.2$ kΩ and $C = 0.01$ μF. The reference voltage is 12 V and the fixed sampling time, while the capacitor is charging, is 8 μs. Find the conversion time for a 7 V input.

12
Review of digital conditioning techniques

12.1 The advantages of digital information

In Chapter 11 we examined several methods of converting an analogue voltage to digital pulses and vice versa. In this chapter we shall be looking further into the handling and manipulation of data in digital form. At this stage the question could be asked as to why bother with digital signal conditioning. The answer is in the advantages that digital representation has over analogue.

Firstly, the transmission of data to, within and from a signal conditioning system can be undertaken more reliably using digital pulses rather than continuously varying analogue signals. This is because the information contained by an analogue signal is embodied in its size and shape and any changes in these qualities during transmission will result in the information received being distorted. The resistive and capacitive elements of a transmission line reduce the amplitude and change the shape of any electrical signal that it is carrying. Further problems arise with the addition of noise voltages to the wanted signal and the drift, with change of temperature, of any processing amplifier gain. In general the analogue representation of transmitted information suffers more so than does digital pulse representation. The discrete digital voltage pulses are all of the same size and width; the information they carry is contained in the way the pulses are arranged into strings of pulses and spaces. The reliable reception of the information is now a matter of detecting only the presence or absence of the pulses; their size and shape on reception are not important.

The second advantage of using digital signals is their ready compatibility with the digital computer, which requires digital inputs rather than analogue. Digital computers are widely used for the control of many commercial and industrial processes. Their speed and work capacity make digital computers ideal for the signal processing and control of systems which comprise many simultaneous interdependent variables. They are also used for improving the linearity of the analogue signals produced by the transducer sensors at the 'front end' of the controlled process systems.

The following sections of this chapter are intended to give an overall review of the fundamental digital techniques involved in signal processing and the control of industrial processes.

12.2 Number systems

The use of digital techniques in process control requires that continuously variable measurement and control information be encoded into current or voltage pulses. The pulses (or digits) themselves are simply two-state (or binary) voltage levels. We describe the two voltage levels as being *high* (H or 1) or *low* (L or 0). Typical nominal voltage values are + 5 V for H and 0 V for L.

12.2.1 Digital word

Since the digital pulse itself can only represent two levels, being either present (1) or absent (0), there needs to be a method of representing a whole range of different analogue values. This is done by stringing pulses together in different patterns of 1 and 0. The string may comprise eight binary levels, for example, 10101010, which could exist in reality as eight parallel wires with + 5 V and 0 V on alternate wires. (The 1 and the 0 are called *bits* since they are *bi*nary digi*ts*.) Alternatively, an 8-*bit* string, or *word,* comprising 11001100 could be represented in serial form by a single wire being strobed sequentially through + 5 V, + 5 V, 0 V, 0 V, + 5 V, + 5 V, 0 V, 0 V. Figure 12.1 shows voltage pulse waveforms which represent these two binary patterns or numbers.

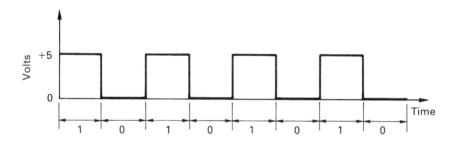

(a)

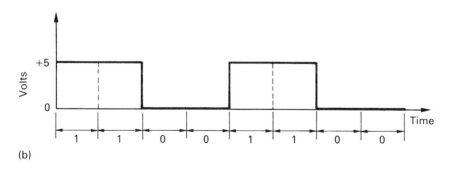

(b)

Figure 12.1 *Voltage waveform representation of digital words. (a) 5 V pulse pattern for the digital word 10101010. (b) 5 V pulse pattern for the digital word 11001100*

There are a few further terms which should be understood. The digital *word length* may be 8, 16, 32 or more bits long. A string of 8 bits is called a *byte*. A byte comprises two 4-bit *nibbles*. The digital word, of whatever length the system designer has decided, can be thought of as a number to the base 2 and the word pulse pattern is really a binary code used to represent different analogue quantities. Perhaps at this point we had better review the main number systems with which we are likely to be involved in our workings with digital signal conditioning.

12.2.2 Decimal numbers

This is the system with which we are all very familiar, but since other number systems follow the same rules we shall remind ourselves of those rules.

The decimal system has *ten* symbols or levels:

0, 1, 2, 3, 4, 5, 6, 7, 8 and 9

The numbers in this group are said to be *units* or in the *house of 10^0* (which is unity, since any number $a^0 = 1$).

For numbers greater than 9 we use the same lot of symbols again but this time prefixing each one with the symbol 1 to 9 as appropriate to form 10 new groups of 10 numbers:

10, 11, 12, 13, 14, 15, 16, 17, 18 and 19

20, 21, 22, 23, 24, 25, 26, 27, 28 and 29

and so on until we reach the tenth and last permissible group of two-digit numbers, namely,

90, 91, 92, 93, 94, 95, 96, 97, 98 and 99

Reading from the left, the first of the two digits is said to be in the *house of tens*; the second digit is in the *house of units*.

For numbers into the hundreds (10^2) and thousands (10^3) and so on, the system is simply extended by repeating the use of further prefix numbers. This produces an endless system of *houses* in increasing powers of *ten*. For example, the decimal (or *denary*) number 4679 can be looked at as comprising:

$$(4 \times 10^3) + (6 \times 10^2) + (7 \times 10^1) + (9 \times 10^0)$$

or

4000   +   600   +   70   +   9

The position of the digit determines its *weighting or the house* in which it resides. The weighting is the power of 10 multiplying factor applied to the digit. Thus, for the complete decimal system, taking into account that we must allow for fractional numbers too, we write the weighting sequence as follows:

$$\ldots 10^3 \ 10^2 \ 10^1 \ 10^0 \quad . \quad 10^{-1} \ 10^{-2} \ 10^{-3} \ldots$$

$$\underleftarrow{\text{whole numbers}} \quad \overrightarrow{\text{fractions}}$$

or

$$\ldots 1000 \ 100 \ 10 \ 1 \quad . \ 1/10 \ 1/100 \ 1/1000 \ldots$$

To indicate that a number is a decimal or denary number, as opposed to one from a system having a different base, we add the subscript:

4679_{10}

12.2.3 Binary numbers

In this system, the same basic rules as for the decimal system apply; the difference lies in there being only two states or levels, 0 and 1, rather than the 10 of the decimal system. So the weightings, or houses, into which the digits are placed are powers of 2 but in just the same pattern as previously and also taking care of binary fractions.

$$\ldots \ 2^3 \ 2^2 \ 2^1 \ 2^0 \quad . \quad 2^{-1} \ 2^{-2} \ 2^{-3} \ \ldots$$

or

$$\ldots \ 8 \quad 4 \quad 2 \quad 1 \quad . \quad 1/2 \ 1/4 \ 1/8 \ \ldots$$

e.g. 1 0 1 0 . 1 0 1 if converted into decimal would be:

$$(1 \times 8) + (1 \times 2) \quad . \quad (1 \times 1/2) + (1 \times 1/8)$$

that is

10.625_{10}

The conversion between the decimal and binary systems is often required and it is recommended that Table 12.1 is memorised.

Table 12.1 *Conversions between binary and decimal*

Binary	Decimal	Binary	Decimal
0000	0	1000	8
0001	1	1001	9
0010	2	1010	10
0011	3	1011	11
0100	4	1100	12
0101	5	1101	13
0110	6	1110	14
0111	7	1111	15

To convert larger decimal numbers to binary the method is to divide the decimal number repeatedly by 2 until zero is the answer but retaining the remainder at each stage of the division:

$19/2 = 9$ rem.1 ← least significant bit (LSB)

$9/2 = 4$ rem. 1

$4/2 = 2$ rem. 0

$2/2 = 1$ rem. 0

$1/2 = 0$ rem. 1 ← most significant bit (MSB)

Therefore,

$\mathbf{19_{10} = 100112_2}$

12.2.4 Negative binary numbers

In the decimal system negative numbers are represented by the normal positive value with a minus sign prefix. This can also be done in the binary system. However, in the binary system it is often useful to let the first digit of the binary number indicate the sign. With an 8-bit computer system, the MSB indicates the sign of the number and the following seven bits the magnitude of the number. Thus, if we were to decide that 1 was the negative sign and 0 the positive, then 10000111 would represent -7_{10} while 01111111 would indicate $+127$. Unfortunately, this simple system does not always produce the expected results from routine mathematical manipulations and so a further refinement was introduced. This entailed coding the last seven bits, which indicate the number magnitude, in a special way using what is called the *two's complement notation*. This is how it works: if we wish to express -8_{10} in 8-bit binary we start by writing down the normal binary for 8_{10}.

$8_{10} = 00001000$

The first digit (MSB) is changed to a 1 to indicate that the number is a negative. Next, the last seven digits are changed to the two's complement to form a new code for the magnitude 8_{10}. The two's complement is formed by *inverting* the last seven digits (write 1 in place of 0 and vice versa) and then adding 0000001, ignoring any carry figure.

	0001000
	0001000
Invert	1110111 − A
Add	0000001 − B

Two's complement $(A + B) = 1111000$

We now replace the negative sign bit at the front of the seven 2's complement bits so

$-8_{10} = \textbf{11111000}$

The above method of producing the two's complement representation of a negative number is the way the 8-bit digital computer does it. An easier way is to take the seven magnitude bits and working from the right (LSB) leave the existing 1 and 0 pattern unchanged up to and including the first 1 encountered; from there, still working to the left, invert the remaining bits. Then add the eighth bit 1 to signify the negative. For example, the two's complement of 0000110 is 1111010; the complete 8-bit negative number would be represented by 11111010.

12.2.5 Hexadecimal numbers

The problem with the binary number system is that while it is the natural language of the digital electronic world, it is very cumbersome and prone to errors when handled by humans. The copying of strings of 8-bit words is tedious enough and even worse when these are extended to 16 bits and above. A useful compromise is the hexadecimal system. As we would expect from our knowledge of the rules as applied to the decimal and binary number systems, the *hex* system counts in houses of powers of 16 and requires 16 symbols to represent its 16 states or levels. In this latter respect, the 16 symbols are the same as the decimal 0 to 9 plus A, B, C, D, E and F to represent 10, 11, 12, 13, 14 and 15 respectively.

The sequence of weightings for the hex system, ignoring the fractional terms for simplicity, is as follows:

$$\ldots 16^3 \quad 16^2 \quad 16^1 \quad 16^0$$

So, the house of units (16^0) takes the symbols 0 to F, that is, decimal 0 to 15 and then repeats the sequence 15 times, each time using successive prefixes, 0 to F, in the house of 16^1. After reaching a count of FF, the house of 16^2 is brought into use and so on.

The number sequence looks like:

0	1	2	3	4	5	6	7	8	9	A	B	C	D	E	F
10	11	12	13	14	15	16	17	18	19	1A	1B	1C	1D	1E	1F
20	21	22	23	24	25	26	27	28	29	2A	2B	2C	2D	2E	2F
30	31 ...														

and so on to

...	F4	F5	F6	F7	F8	F9	FA	FB	FC	FD	FE	FF
110	111	112	113 ... and so on									

In order to distinguish the hex series of numbers from the binary and decimal, the suffix H or (hex) or the subscript 16 is used. For example,

$111H = 111(\text{hex}) = 111_{16}$

If we convert this hex number into decimal it becomes

$$(1 \times 16^2) + (1 \times 16^1) + (1 \times 16^0) = 273_{10}$$

Similarly, BH, 3CH and FADEH would respectively convert into 11_{10}, 60_{10} and 64222_{10}.

To convert decimal into hex, the decimal number is repeatedly divided by 16, keeping the remainder at each step. Thus, it can be shown that $26_{10} = 1AH$, $202_{10} = CAH$ and $452_{10} = 1C4H$.

To convert binary into hex, starting from the LSB divide the binary number into groups of 4 bits. It may be necessary to add zeros to the front of the left-most group to make it up to 4 bits. For example, to convert 1111111010011_2 into hex, start by dividing it into groups of 4 bits, noting that the left group will need the addition of two zeros to make up the 4 bits, so:

0011 1111 1010 0011

Now read the binary value of each group in decimal but write it in hex so:

3   F   A   3 = 3FA3H

To convert from hex to binary is simply the reverse of the above.

12.2.6 Binary coded decimal (BCD)

Finally we shall mention the BCD system which is much used in digital electronics. This is a method whereby the individual decimal figures are converted into a group of 4 bits. For example, to convert 26_{10} into BCD, we write the 2 as 0010 and the 6 as 0110 and the complete conversion becomes 00100110_2. The procedure is reversed for converting from BCD to decimal.

12.3 Boolean algebra and logic gates

Boolean algebra is a mathematical method of defining the operation of a process the control of which can be achieved by a system of switched two-state devices. The devices are either ON or OFF, HIGH or LOW, ENABLED or DISABLED, RUNNING or STOPPED, CLOSED or OPEN; that is, in general, at a logic 1 or logic 0. The logic switching is undertaken by combinations of electronic logic blocks called *logic gates* and before we proceed we should review the techniques involved with both Boolean logic algebra and logic gates.

12.3.1 Boolean algebra

The logic manipulation of the two states, 0 and 1, is by four logic operations as follows:

- *Equality* (symbol =). If two quantities, A and B, are equal we mean that whatever the logic state of one, the other is the same. This is expressed as A = B so if A = 1, then B = 1; if A = 0, then B = 0.
- *Non-equality or complement* (the symbol is a line or bar above the variable, e.g. $\overline{A}$. The interpretation of this is that if A = 1 then $\overline{A}$ does NOT EQUAL 1 and therefore, since we are operating a two-state system, $\overline{A}$ can only be 0. Similarly, if A = 0, then $\overline{A}$ = 1.
- *AND* (the symbol is a dot). This logic operation is used to signify that two or more variables must co-exist all at a logic 1 in order that their joint outcome is also a logic 1. A.B.C.D = F means that A, B, C and D must all equal 1 in order that F = 1. Should any one of the four variables be a 0 then F = 0.
- *OR* (symbol +). The logic of this operation is that if two or more variables co-exist to produce an outcome, then that outcome will be a 1 if any of the variables is a 1.

A process comprising two-state switching operations can be modelled mathematically using the above logic operations. For example,

$$A + B.C = F$$ means that F will be a 1 provided that A = 1 or B and C are both 1.

$$\overline{A} + C.\overline{D} = F$$ means that F will be a 1 provided that A = 0 or C and D are together 1 and 0 respectively.

   Some processes produce rather long logic expressions and there are laws and rules for simplifying these. The detailed explanation of the mathematical manipulation involved is outside of the scope of this text but it is worth learning bearing in mind that the implementation of the final Boolean expression is by using logic gates. These are commercially available integrated circuits and the simpler the logic expression the fewer the number of gates needed.

12.3.2 Logic gates

These are self-contained logic integrated circuits which are connected together to build the control or conditioning circuits required to fulfil a designed logical outcome. Each gate has its own logic symbol and these, together with the relevant Boolean expression for the output, are shown in Figure 12.2.

12.4 Boolean algebra – a practical application

Let us see how this could work in practice. Suppose we have a process which uses a rotary pump to pass hot fluid around a system. The general arrangement

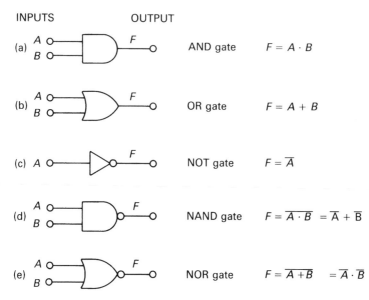

Figure 12.2 *Symbols for a two-input logic gate*

is shown by Figure 12.3. The pump outlet pressure, A, the pump angular velocity, N, and the fluid temperature, T, are monitored by transducer–comparator combinations to indicate pre-set high and low limits. The comparators produce a logic 1 at a high limit and a logic 0 at a low. Further suppose that there is a requirement for an alarm to be raised should combinations of these high and low limits occur as follows:

- low pressure plus high pump speed $(\overline{A}N)$, or
- high pressure plus high temperature (AT), or
- high pressure plus high temperature plus low speed $(AT\overline{N})$.

This full requirement can be expressed in the Boolean equation

$$F = \overline{A}N + AT + AT\overline{N}$$

(Note how the Boolean operation AND 'dot' may be omitted; $A.T$ can be written as AT.)

We can use the logic symbols shown in Figure 12.2 to design a logic diagram which will produce the required control function. This is done quite simply by first making a visual inspection of the terms comprising F and then working back to the individual sources of A, N and T, putting in the necessary logic gates to produce the required logic paths. There are several solutions, all of which will work. One such solution is the logic diagram shown in Figure 12.4.

Figure 12.4 uses a mixture of AND, OR and NOT gates. It is possible to use only one type of gate throughout the whole logic diagram. Of course there are

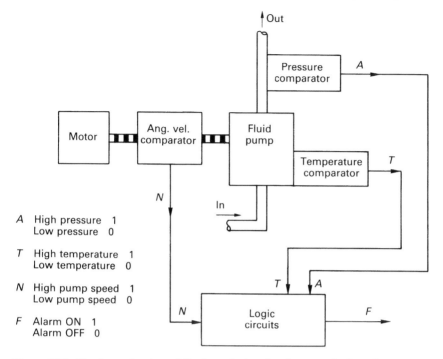

Figure 12.3 *Simple application of Boolean algebra for the control of an alarm*

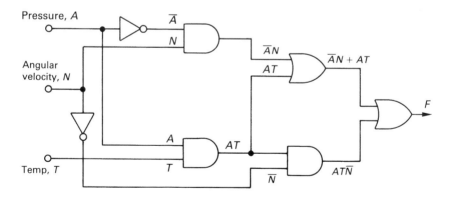

Figure 12.4 *One possible logic diagram realisation of the Boolean equation*
$$F = \overline{A}N + AT + AT\overline{N}$$

more gates involved but that is of little consequence with the small, low cost, multi-gate integrated circuits which are now available. The NAND gate is as easy to manufacture as any and is much used as the standard single logic gate in many control functions.

12.5 Programmable logic controllers (PLCs)

PLCs are devices which are particularly suited to control systems which use binary logic techniques. They are regarded by some as a development of the older relay sequence controller and this statement gives us a lead into the understanding of how PLCs function. First we must appreciate the type of control situation best suited to the PLC and then how it works.

12.5.1 Discrete state process control

One way of controlling the variable temperature of an oven is continuously to adjust the heat input. A different way is to set an upper and lower temperature and simply switch the heat ON when the temperature reaches the lower limit and OFF when it reaches the high limit. This method of control is now one of sequential switching rather than continuous adjustment. It is known as *discrete state process control* because all the measurements taken from and all the controlling inputs to the process are in discrete binary states. The control of the process is effected through a sequence of *events* which, in the case of our example of oven temperature control, would be:

- temperature low, heater off;
- temperature low, heater on;
- temperature high, heater on;
- temperature high, heater off.

Each event remains unchanged until the temperature changes from one discrete limit to another; there is no intermediate adjustment.

In general then, the discrete state control of a process means that measurements are *high* or *low*, *above* or *below*, motors and switches are *on* or *off*, valves are *open* or *closed*.

12.5.2 Example application

Suppose that a goods scissor lift is made to rise by a motor driven hydraulic pump which supplies hydraulic fluid under pressure to a ram which operates the scissor mechanism. When the motor stops running, the platform stops lifting and stays at its present height. The platform is lowered by opening a hydraulic solenoid valve which releases the high pressure fluid from the ram, which falls under gravity.

The mechanical layout is shown in Figure 12.5. The sequence of events which control the operation of the lift could be documented using the binary statement as follows:

(US)(DS)(PD)(PU)(MC)(HS)

Each event in the operational sequence would produce a 6-digit binary input word and a 2-digit output signal as shown below:

	Event	*Input*	*Output*
1.	Load goods on platform	001000	00
2.	Set lift control to UP	101010	10
3.	Test for platform fully UP		
	(a) NO – retain output	101010	10
	(b) YES – stop lift motor	100100	00
4.	Unload platform		
5.	Set lift control to DOWN	010101	01
6.	Test for platform fully DOWN		
	(a) NO – retain output	010101	01
	(b) YES – close hydraulic valve	011000	00
7.	Go back to 1		

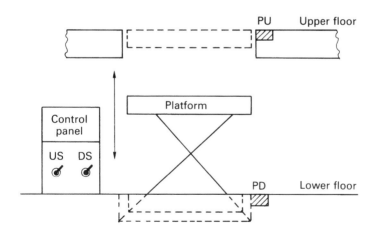

MC Hydraulic pump drive motor: ON = 1, OFF = 0
HS Hydraulic solenoid valve: OPEN = 1, CLOSED = 0
PU Limit switch: FULLY UP = 1, NOT UP = 0
PD Limit switch: FULLY DOWN = 1, NOT DOWN = 0
US UP control: UP = 1, OFF = 0
DS DOWN control: DOWN = 1, OFF = 0

Figure 12.5 *Mechanical layout and control logic for scissor lift application*

This is the approach which would best suit computer control but there is an alternative control method using a system of relays and switches. A simplistic wiring diagram for the scissor lift shown in Figure 12.5 is shown in Figure 12.6 the operation of which is now explained.

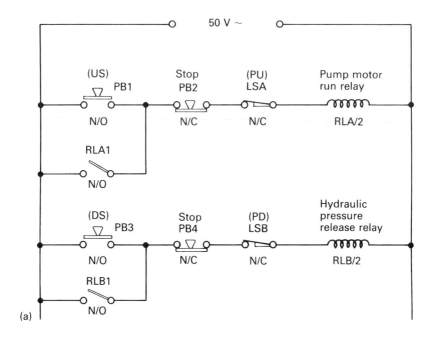

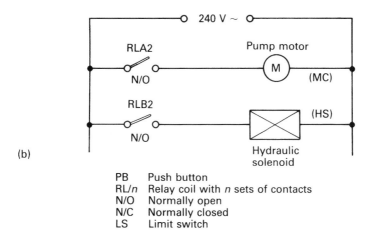

PB	Push button
RL/*n*	Relay coil with *n* sets of contacts
N/O	Normally open
N/C	Normally closed
LS	Limit switch

Figure 12.6 *Wiring diagram for the scissor lift shown in Figure 12.5*

The total arrangement comprises a low voltage control circuit and a normal mains voltage circuit for powering the pump motor. To cause the platform to rise, the start button, PB1, is pressed, allowing current to flow through PB2 and LSA to energise the relay RLA/2. Contacts RLA1 and RLA2 close respectively to short out PB1 and start the pump motor running. The latching effect of RLA1 causes the pump motor to continue running after PB1 has been released. Stop button PB2 is available for emergency use to break the electrical supply to the pump motor run relay. When the lift platform has reached the correct height, it opens LSA, effectively breaking the current supply to the pump motor, and so stops the platform. A similar procedure is followed in order to lower the platform. PB3 is pressed and RLB/2 is energised. The contacts RLB1 short out PB3 and contacts RLB2 energise the hydraulic fluid pressure release solenoid, allowing the platform to fall under its own weight.

We could easily add more control functions to Figure 12.6. For example, a 'power on' indicator light may be considered useful and a powered belt conveyor mounted on the platform would ease the task of unloading the platform. Figure 12.7 shows the additional 'rungs' needed to provide these extra services. The 'ladder-like' circuit contains the original rungs 1 and 2 from Figure 12.6 and rungs 3 and 4 which provide the additional controls. The indicator light in rung 3 is ON whenever power is connected to the circuit and the belt conveyor will run as long as PB5 is depressed.

It should be appreciated that Figure 12.7 is not a truly practical circuit. For example, there would need to be safety interlocks to ensure that the platform DOWN action could not be initiated while the system was in the UP mode. Also, it is undesirable that the belt motor can be made to run unless the platform is either fully up or down.

Figure 12.7 is sometimes called a ladder diagram and represents the total arrangement of relays, switches and contacts which in effect constitutes a special sequential relay controller for a particular process. The components are all 'hard-wired' together in such a manner so as to undertake the particular control function for which it has been designed. Any changes in the state of a switch, contactor or relay has an immediate effect upon the circuit. Further, any modifications to how the circuit works require work on circuit hardware.

Some of the contacts in Figure 12.7 are simply 'control contacts', that is they are not handling the final switching operation to route driving current to a motor or solenoid. Typical of these are the latching contacts RLA1, RLA1 and RLC1. A new method of controlling processes is by the use of a *programmable logic controller*. This eliminates the necessity for some of the control contacts and relays by the alternative use of software. With the development of new type solid state electronic switches such as the silicon controlled rectifier (SCR) it is now possible to have high power switching to motors and heaters initiated by low level commands from computers. The basic functional block diagram of a programmable logic controller is shown in Figure 12.8.

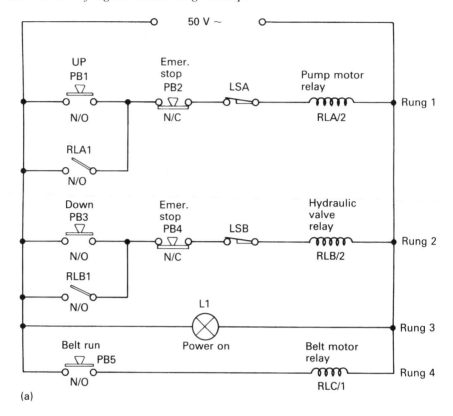

(a)

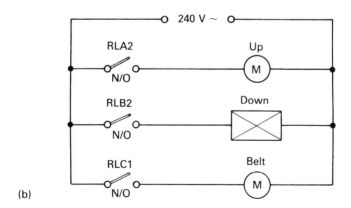

(b)

Figure 12.7 *Rungs 3 and 4 provide added controls. (a) The control circuit; (b) the power circuit*

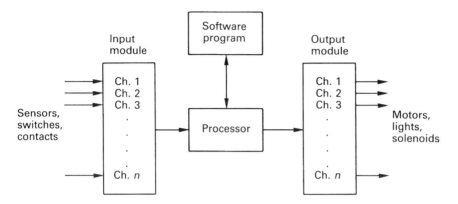

Figure 12.8 *Block diagram of a programmable logic controller*

12.5.3 PLC block diagram

Figure 12.8 shows the functional block diagram of a PLC. The processor at the centre of the system receives binary signals from the input module, processes them in accordance with the sequence of switching (or process control program) stored in the software program and passes the conditioned signals through the output module to the motors and valves of the process.

The processor executes the prescribed sequence of switching to perform the operations specified by the ladder diagram or by a set of Boolean equations. The processor undertakes the arithmetic and logic operations on the varying input data and produces the correct combinations of output states. Because the processor is virtually a computer, it has to accept the input data in serial form and needs to 'look' at each input channel in rapid sequence. In this respect it differs from the old system of hard-wired relays in that it does not have the same instantaneous response to an input change. Any change on an input channel is not noticed until that channel is *scanned* in its turn under the control of the processor.

The input module sequentially scans the outputs of input devices which are attached to it and translates each particular voltage state into a binary 1 or 0 for recognition by the processor. Figure 12.9 shows typical connections which could be used for the scissor lift control circuit shown in Figure 12.7. When in their logic 1 state, the various push buttons and limit switches connect the full a.c. supply voltage into their allotted input module channels; in their logic 0 state, zero volts are connected. For example, the a.c. supply could be at 110 V, 50 Hz for a logic 1 and the input module would translate this to $+5$ V d.c. for the processor input. Some types of input modules have a light-emitting diode (LED) indicator light for each channel to tell if the channel is ON or OFF.

The input module receives the processed binary output from the processor and routes this to the correct output channel at the correct power to activate the output device. Such an arrangement is shown in Figure 12.10. Once again, output channel ON or OFF indicator lights may be fitted.

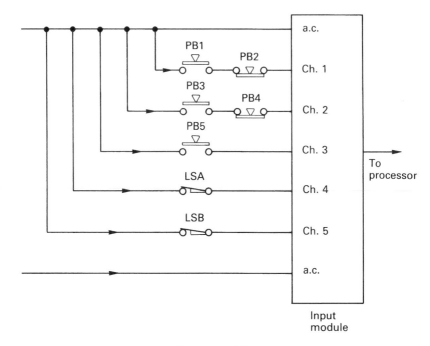

Figure 12.9 *Connections to a PLC input module*

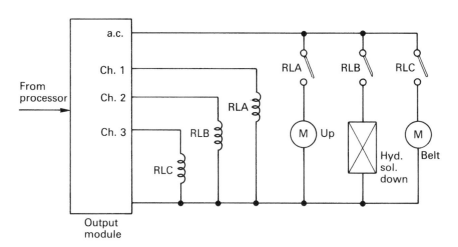

Figure 12.10 *Connections to a PLC output module*

12.5.4 PLC operation

This is not a continuous monitoring and adjustment mode but rather one where all inputs and outputs are scanned and updated one channel at a time. If an input has changed, then the processor evaluates the effects, one 'ladder rung' at a time, and then updates the outputs accordingly. The speed of the processor clock decides the total time needed to implement a complete scan–process–adjust cycle but typically each channel would be read or updated at least 50 times per second.

12.5.5 Programming

The program initially is fed into the programmable controller from an external plug-in programming unit. The programming unit has a liquid crystal display and the program elements, entered through a coded keyboard, are shown as rungs of the appropriate ladder diagram. The programming unit also can be used for changing existing software programs to implement changes in the controller's operation without the necessity of changing the hard-wired circuitry.

12.6 Buses and tri-state buffers

We have seen in the previous section how the processor is used to process the signal data it receives and then to send that information to other parts of the system. In order that we can appreciate how a digital computer can receive or send binary 'words' we need to be aware of two important devices.

12.6.1 Buses

These are no more than eight parallel copper conductors in a single ribbon cable which can simultaneously carry the eight separate voltages which constitute an 8-bit word. Since these 'eight-lane highways' are required for use by more than one send and one receive device, they are sometimes referred to as omnibuses or simply *buses*. When a set of data has been successfully transferred it is important that the send and receive devices can both be disconnected from the common bus, so freeing it for other users. This means that the output of the digital devices connected to the common bus must be capable of switching not only to a logic 1 or logic 0 but additionally to a *high impedance* state which is a virtual disconnection. This is the purpose of the tri-state buffer.

12.6.2 Tri-state buffer

This device acts like a three-way switch. In the first position it places its high output, 1, on the bus, in the second position its low output, 0, and in the third position it is disconnected. The normal situation is that while each send and receive device is physically connected to the common bus through its own tri-state buffer, it is electrically disconnected from it because the tri-state buffer is normally in its high impedance mode. When the tri-state buffer is *enabled* by a control signal to its enabling terminal, the high or low state of the device output is connected to the bus. On completion of the data transfer, the tri-state buffer is disabled and the device is isolated again.

12.7 Data acquisition systems

Because computers used for the control of slowly changing analogue processes operate so quickly, the two must be carefully linked together. The computer can compare a pair of numbers or add them in a few microseconds whereas the control data being extracted from the process sensors may be taking minutes or even hours before a change is noticed. The usual way of linking or *interfacing* the fast computer and the slow process is through a *data acquisition system* (DAS). The computer takes frequent sample readings of the slowly changing process parameters and after processing them, updates the control outputs as necessary. A typical data processing system is shown by the block diagram in Figure 12.11. The various blocks contain circuits which are readily available commercially in integrated circuit form and can be interconnected to form the DAS.

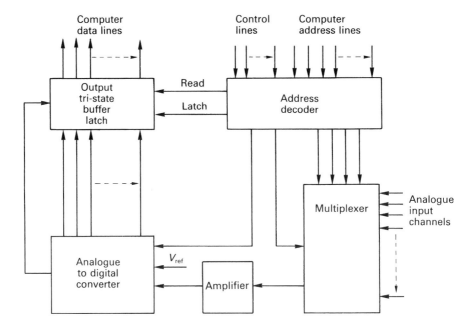

Figure 12.11 *Data acquisition system*

Address decoder. This device is connected to the computer control and address buses. The address may comprise 16 or more bits depending upon the type of computer. We shall assume a computer that uses 8-bit data words in which case the address word usually will 16 bits long. The 16 address bits received by the address decoder are processed into a 4-bit output which is passed to the multiplexer for it to select a specific one of 16 analogue input channels for sampling.

Multiplexer. Figure 12.12 shows how this device operates. It receives the 4-bit code from the address decoder which, in the case of our example, would be 0101. This actuates a solid state electronic switch which effectively connects analogue input number 5 to the multiplexer output. In order to sample the 16th analogue input channel, the address decoder would feed 1111 to the multiplexer and so select number 15, and so on.

Amplifier. The analogue-to-digital converter is a device which operates best if its input analogue voltage lies within a range prescribed by its manufacturer. If this range were 0 V to 2.5 V and the analogue input voltages were in the region of a few millivolts, then there is a case for having an adjustable gain amplifier fitted within the DAS. External pre-amplification on specific channels may also be needed should there be a large spread of analogue input voltages from the process sensors.

Analogue-to-digital converter (ADC). This is an obvious requirement to convert the analogue input signal into a form which the digital computer can handle. The operation of ADCs is dealt with in some detail in Chapter 11.

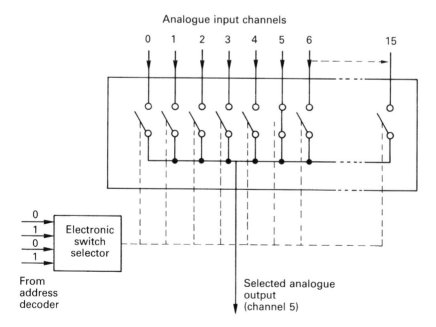

Figure 12.12 *A 16-channel analogue multiplexer*

Output buffer. This holds the 8-bit data signals which the ADC has produced from the analogue inputs. The data are held in the buffer until the computer has read them or until they require updating.

Control. This function is not shown in Figure 12.13 as a specific block but is implemented through the control lines from the computer. The control function is vital in keeping the timing of the different DAS operations in the correct sequence.

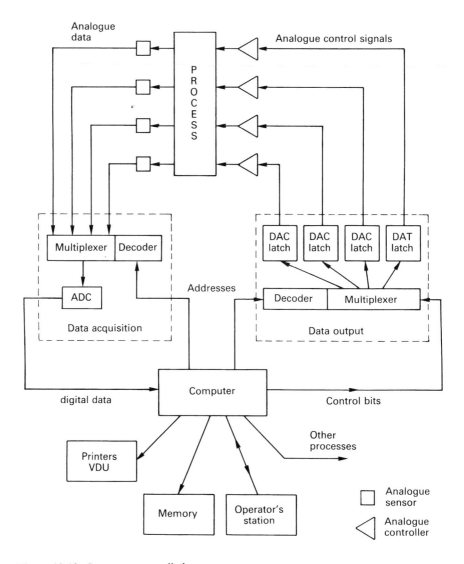

Figure 12.13 *Computer controlled process*

12.8 Sampling and holding

In order to convert a continuous analogue voltage into a series of discrete pulses we pass a sample of the former through an ADC. The pulse pattern produced by the ADC is indicative of the amplitude of the sampled voltage. It is often desirable that the sample voltage is made available for conditioning some time after it was taken. This calls for the use of a *sample and hold circuit*. Since this subject has been dealt with in Sections 7.9 and 11.2, the reader is referred to this earlier text for further information.

12.9 Encoding and decoding

In simplistic terms, *encoding* means the conditioning of an analogue quantity into a digital word (binary number) which is coded to indicate the analogue magnitude. *Decoding* usually entails translating one digital code into another or to an analogue quantity.

12.10 Microcomputer control

Figure 12.13 shows the block diagram of a typical computer controlled process. The diagram contains nothing which has not been mentioned previously and merely serves as a meeting point. Doubtless most readers will find the operation of the system self-evident. Clearly, the left side is concerned with the gathering of measurement data from the process analogue sensors and its conversion into digital form for passing to the computer for conditioning. The program contained within the computer examines the received measurement data with pre-set numbers and from this produces the digital control signals. On the right side of the diagram, these are converted to analogue and passed to the process analogue controllers.

Unlike the dedicated processor used with the discrete two-state control process shown in Figure 12.8, this computer uses normal computer programming language. It will be recalled that the dedicated programming of the processor system required the use of special symbols and a ladder diagram.

12.11 Digital filters

In Chapter 9 we discussed the action and use of analogue filters but made no specific mention of digital filters which are designed for use with pulsed inputs. This section is intended to serve only as a very brief, perhaps simplistic, introduction to the operation of digital filters. A deeper study would require the use of analytical mathematics outside the scope of this text.

While the output from a process is clearly dependent upon the present input there are situations where the output may also be dependent upon previous inputs and previous outputs. This is particularly the case when the input, and therefore the output, is a series of varying amplitude pulses each being applied to the process before the effects of the previous pulses have completely subsided. A digital filter can be regarded as such a process. A typical application for a digital filter is where an analogue filter would be difficult to make work properly; for example, in separating one very low frequency waveform from another. A digital filter can be made to operate with signal frequencies as low as a fraction of 1 Hz.

The digital filter works by rapidly sampling a low frequency analogue wave and producing a train of discrete pulses, spaced equally in time, but each having a magnitude proportional to the instantaneous analogue signal amplitude at the time the sample was recorded. The sampling rate must be at least twice the analogue signal frequency; otherwise, problems of *aliasing* can occur. (This is a problem associated with the conversion of the filtered digital signal back into analogue form when the output is of the incorrect lower alias frequency.)

Now, if a single pulse (or impulse) is applied to the input of a digital filter, which is an electronic circuit having inductance and capacitance, the circuit responds by 'vibrating' electrically or 'ringing'. If a second pulse is applied, possibly of a different amplitude to the first, before the ringing has ceased then the effects of the two pulses are superimposed. Multiply this action by the combined effects of a whole series of input pulses and an idea of the complexity of the filter output can be imagined. Past work in this field has shown that the output from some digital filters can be represented mathematically by the following equation:

$$y_k = x_k + A y_{k-1} + B y_{k-2}$$

where A and B are constants, y_k is the present output, x_k is the present input, y_{k-1} is the previous output and y_{k-2} is the output before that.

This particular equation, or *transfer function*, represents what is known as a *recursive* filter because its output depends upon previous outputs as well as the present and previous inputs. A system or filter is said to be *non-recursive* if its output depends only upon the inputs previous and present.

A discrete sinusoidal waveform is the series of sampled pulses taken at regular time intervals from a normal analogue sine wave; in other words, the amplitude of the pulse train varies sinusoidally. If we apply this discrete sinusoidal wave to a *linear discrete system,* which is what a digital filter effectively is, then the output will be sinusoidal at the same frequency but possibly of different amplitude and with a phase shift. This response can be used to determine the frequency response of the filter, that is, where its frequency break point and -3 dB fall in gain occur. However, for those familiar with the mathematical manipulations using what is called the *z*-transform, as opposed to Laplace or Fourier Transforms used with analogue signals, it is often easier to use the transfer function. This is given values for A and B and the output calculated and plotted to obtain the required low-, high- or band-pass filter response.

The filter itself can be made using hardware, possibly in the form of a dedicated system of integrated circuits, or incorporated in the software program of a normal digital computer. The first system is rapid in action, as required for an on-line real time filter, but can be expensive. While the second method is slower, it is also cheaper, and if the signal to be filtered can be stored, that is, it is not a *real time* signal, the slowness is of no consequence.

Exercises

12.1 Convert to binary:
(a) 14_{10}, (b) 96_{10}, (c) 247_{10}, (d) 1025_{10}.

12.2 Convert to binary: (a) 0.5_{10}, (b) 0.05_{10}, (c) 0.975_{10} (d) 0.83_{10}.

12.3 Convert to decimal: (a) 111100_2, (b) 11001101_2, (c) 100010001_2.

12.4 Convert to hexadecimal: (a) 10111_2, (b) 1110111_2, (c) 111100100_2.

12.5 Convert to binary: (a) 16H, (b) A6FH, (c) B3F2H.

12.6 Convert to decimal: (a) 27_{16}, (b) $2AA_{16}$, (c) 1111_{16}.

Appendix 1

Operational amplifier data sheets

Texas Instruments of Manton Lane, Bedford, England, have given their kind permission for the inclusion of the following data sheets in this book:

- Operational Amplifiers – Glossary
- Operational Amplifiers – LM108, LM108A, LM208, LM208A, LM308, LM308A.

These data sheets are reproduced exactly as they appear in the *Texas Instrument Linear Circuits Handbook*.

OPERATIONAL AMPLIFIERS GLOSSARY

Input Offset Voltage (V_{IO})

The d-c voltage that must be applied between the input terminals to force the quiescent d-c output voltage to zero or other level, if specified.

Average Temperature Coefficient of Input Offset Voltage (α_{VIO})

The ratio of the change in input offset voltage to the change in free-air temperature. This is an average value for the specified temperature range.

$$\alpha_{VIO} = \left[\frac{(V_{IO} @ T_{A(1)}) - (V_{IO} @ T_{A(2)})}{T_{A(1)} - T_{A(2)}} \right] \text{ where } T_{A(1)} \text{ and } T_{A(2)} \text{ are the specified temperature extremes.}$$

Input Offset Current (I_{IO})

The difference between the currents into the two input terminals with the output at zero volts.

Average Temperature Coefficient of Input Offset Current (α_{IIO})

The ratio of the change in input offset current to the change in free-air temperature. This is an average value for the specified temperature range.

$$\alpha_{IIO} = \left[\frac{(I_{IO} @ T_{A(1)}) - (I_{IO} @ T_{A(2)})}{T_{A(1)} - T_{A(2)}} \right] \text{ where } T_{A(1)} \text{ and } T_{A(2)} \text{ are the specified temperature extremes.}$$

Input Bias Current (I_{IB})

The average of the currents into the two input terminals with the output at zero volts.

Common-Mode Input Voltage (V_{IC})

The average of the two input voltages.

Common-Mode Input Voltage Range (V_{ICR})

The range of common-mode input voltage that if exceeded will cause the amplifier to cease functioning properly.

Differential Input Voltage (V_{ID})

The voltage at the noninverting input with respect to the inverting input.

Maximum Peak Output Voltage Swing (V_{OM})

The maximum positive or negative peak output voltage that can be obtained without waveform clipping when the quiescent d-c output voltage is zero.

Maximum Peak-to-Peak Output Voltage Swing (V_{OPP})

The maximum peak-to-peak output voltage that can be obtained without waveform clipping when the quiescent d-c output voltage is zero.

Large-Signal Voltage Amplification (A_V)

The ratio of the peak-to-peak output voltage swing to the change in input voltage required to drive the output.

Differential Voltage Amplification (A_{VD})

The ratio of the change in output voltage to the change in differential input voltage producing it.

OPERATIONAL AMPLIFIERS
GLOSSARY

___

Maximum-Output-Swing Bandwidth (B_{OM})

The range of frequencies within which the maximum output voltage swing is above a specified value.

Unity-Gain Bandwidth (B_1)

The range of frequencies within which the open-loop voltage amplification is greater than unity.

Phase Margin (ϕ_m)

The absolute value of the open-loop phase shift between the output and the inverting input at the frequency at which the modulus of the open-loop amplification is unity.

Gain Margin (A_m)

The reciprocal of the open-loop voltage amplification at the lowest frequency at which the open-loop phase shift is such that the output is in phase with the inverting input.

Input Resistance (r_i)

The resistance between the input terminals with either input grounded.

Differential Input Resistance (r_{id})

The small-signal resistance between the two ungrounded input terminals.

Output Resistance (r_o)

The resistance between the output terminal and ground.

Input Capacitance (C_i)

The capacitance between the input terminals with either input grounded.

Common-Mode Input Impedance (z_{ic})

The parallel sum of the small-signal impedance between each input terminal and ground.

Output Impedance (z_o)

The small-signal impedance between the output terminal and ground.

Common-Mode Rejection Ratio (k_{CMR}, CMRR)

The ratio of differential voltage amplification to common-mode voltage amplification.
NOTE: This is measured by determining the ratio of a change in input common-mode voltage to the resulting change in input offset voltage.

Supply Voltage Sensitivity (k_{SVS}, $\Delta V_{IO}/\Delta V_{CC}$)

The absolute value of the ratio of the change in input offset voltage to the change in supply voltages producing it.
NOTES: 1. Unless otherwise noted, both supply voltages are varied symmetrically.
          2. This is the reciprocal of supply voltage rejection ratio.

Supply Voltage Rejection Ratio (k_{SVR}, $\Delta V_{CC}/\Delta V_{IO}$)

The absolute value of the ratio of the change in supply voltages to the change in input offset voltage.
NOTES: 1. Unless otherwise noted, both supply voltages are varied symmetrically.
          2. This is the reciprocal of supply voltage sensitivity.

OPERATIONAL AMPLIFIERS
GLOSSARY

Equivalent Input Noise Voltage (V_n)

The voltage of an ideal voltage source (having an internal impedance equal to zero) in series with the input terminals of the device that represents the part of the internally generated noise that can properly be represented by a voltage source.

Equivalent Input Noise Current (I_n)

The current of an ideal current source (having an internal impedance equal to infinity) in parallel with the input terminals of the device that represents the part of the internally generated noise that can properly be represented by a current source.

Average Noise Figure ($\overline{F}$)

The ratio of (1) the total output noise power within a designated output frequency band when the noise temperature of the input termination(s) is at the reference noise temperature, T_0, at all frequencies to (2) that part of (1) caused by the noise temperature of the designated signal-input termination within a designated signal-input frequency band.

Short-Circuit Output Current (I_{OS})

The maximum output current available from the amplifier with the output shorted to ground, to either supply, or to a specified point.

Supply Current (I_{CC})

The current into the V_{CC} or V_{CC+} terminal of an integrated circuit.

Total Power Dissipation (P_D)

The total d-c power supplied to the device less any power delivered from the device to a load. NOTE: At no load: $P_D = V_{CC+} \cdot I_{CC+} + V_{CC-} \cdot I_{CC-}$.

Crosstalk Attenuation (V_{o1}/V_{o2})

The ratio of the change in output voltage of a driven channel to the resulting change in output voltage of another channel.

Rise Time (t_r)

The time required for an output voltage step to change from 10% to 90% of its final value.

Total Response Time (Settling Time) (t_{tot})

The time between a step-function change of the input signal level and the instant at which the magnitude of the output signal reaches for the last time a specified level range ($\pm \epsilon$) containing the final output signal level.

Overshoot Factor

The ratio of (1) the largest deviation of the output signal value from its final steady-state value after a step-function change of the input signal, to (2) the absolute value of the difference between the steady-state output signal values before and after the step-function change of the input signal.

Slew Rate (SR)

The average time rate of change of the closed-loop amplifier output voltage for a step-signal input.

LM108, LM108A, LM208, LM208A, LM308, LM308A
OPERATIONAL AMPLIFIERS

D2808, OCTOBER 1983 – REVISED FEBRUARY 1991

- Input Offset Current . . . 200 pA Max at 25°C for LM108, LM108A, LM208, LM208A

- Input Bias Current . . . 2 nA Max at 25°C for LM108, LM108A, LM208, LM208A

- Supply Current . . . 600 μA Max at 25°C for LM108, LM108A, LM208, LM208A

- Input Offset Voltage . . . 500 μV Max at 25°C for LM108A, LM208A, LM308A

- Offset Voltage Temperature Coefficient . . . 5 μV/°C Max for LM108A, LM208A, LM308A

- Supply Voltage Range . . . ±2 V to ±18 V

- Applications:
  - Integrators
  - Transducer Amplifiers
  - Analog Memories
  - Light Meters

- Designed To Be Interchangeable With National LM108 Series and Linear Technology LM108 Series

D, JG, OR P PACKAGE
(TOP VIEW)

```
         ___ ___
COMP1 [ 1  U  8 ] COMP2
 IN – [ 2     7 ] V_CC +
 IN + [ 3     6 ] OUT
 V_CC – [ 4   5 ] NC
```

L PACKAGE
(TOP VIEW)

```
              COMP2
  COMP1    _____  V_CC +
      ( 1 )( 8 )( 7 )
  IN – ( 2 )       ( 6 ) OUT
      ( 3 )( 4 )( 5 )
  IN + (     )       NC
           V_CC –
```

NC – No internal connection
Pin 4 (L package) is in electrical contact with the case.

symbol

description

The LM108 series of precision operational amplifiers is particularly well-suited for high-source-impedance applications requiring low input offset and bias currents as well as low power dissipation. Unlike FET input amplifiers, the input offset and bias currents of the LM108 series do not vary significantly with temperature. Advanced design, processing, and testing techniques make this series a superior choice over previous devices. For applications requiring higher performance, see the LT1008 and LT1012.

The LM108 and LM108A are characterized for operation over the full military temperature range of –55°C to 125°C. The LM208 and LM208A are characterized for operation from –40°C to 105°C. The LM308 and LM308A are characterized for operation from 0°C to 70°C.

AVAILABLE OPTIONS

T_A	V_IO max AT 25°C	PACKAGE			
		SMALL OUTLINE (D)	CERAMIC DIP (JG)	METAL CAN (L)	PLASTIC DIP (P)
0°C to 70°C	0.5 mV	LM308AD	———	———	LM308AP
	7.5 mV	LM308D	———	———	LM308P
– 40°C to 105°C	0.5 mV	LM208AD	———	———	LM208AP
	2 mV	LM208D	———	———	LM208P
– 55°C to 125°C	0.5 mV	LM108AD	LM108AJG	LM108AL	LM108AP
	2 mV	LM108D	LM108JG	LM108L	LM108P

The D package is available taped and reeled. Add the suffix R to the device type (e.g., LM308ADR).

LM108, LM108A, LM208, LM208A, LM308, LM308A
OPERATIONAL AMPLIFIERS

---

schematic

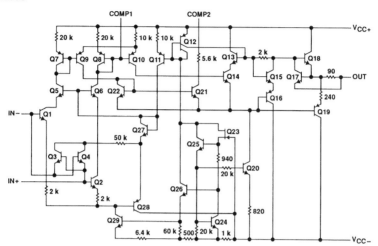

All resistor values shown are nominal and in ohms.

absolute maximum ratings over operating free-air temperature range (unless otherwise noted)

Supply voltage, V_{CC+} (see Note 1): LM108, LM108A, LM208, LM208A . . . . 20 V

LM308, LM308A . . . . 18 V

Supply voltage, V_{CC-} (see Note 1): LM108, LM108A, LM208, LM208A . . . . −20 V

LM308, LM308A . . . . −18 V

Input voltage range, V_I (see Note 2) . . . . ±15 V

Differential input current (see Notes 3 and 4) . . . . ±10 mA

Duration of output short-circuit at (or below) 25°C (see Note 5) . . . . unlimited

Operating free-air temperature range, T_A: LM108, LM108A . . . . −55°C to 125°C

LM208, LM208A . . . . −40°C to 105°C

LM308, LM308A . . . . 0°C to 70°C

Storage temperature range . . . . −65°C to 150°C

Lead temperature 1,6 mm (1/16 inch) from case for 10 seconds: D or P package . . . . 260°C

Lead temperature 1,6 mm (1/16 inch) from case for 10 seconds: JG or L package . . . . 300°C

NOTES: 1. All voltage values, unless otherwise noted, are with respect to the midpoint between V_{CC+} and V_{CC-}.
2. The magnitude of the input voltage must never exceed the magnitude of the supply voltage or 15 V, whichever is less.
3. The inputs are shunted with two opposite-facing base-emitter diodes for over-voltage protection. Therefore, excessive current will flow if a differential input voltage in excess of approximately 1 V is applied between the inputs unless some limiting resistance is used.
4. Differential voltages are at the noninverting input terminal with respect to the inverting input terminal.
5. The output may be shorted to ground or either power supply.

LM108, LM108A, LM208, LM208A, LM308, LM308A
OPERATIONAL AMPLIFIERS

recommended operating conditions

	LM108, LM108A		LM208, LM208A		LM308, LM308A		UNIT
	MIN NOM	MAX	MIN NOM	MAX	MIN NOM	MAX	
Supply voltage, V_{CC+}	5	20	5	20	5	20	V
Supply voltage, V_{CC-}	-5	-20	-5	-20	-5	-20	V
Operating free-air temperature, T_A	-55	125	-40	85	0	70	°C

electrical characteristics at specified free-air temperature, $V_{CC\pm} = \pm 5\,V$ to $\pm 20\,V$ (unless otherwise noted)

PARAMETER		TEST CONDITIONS	$T_A{}^\dagger$	LM108A, LM208A			LM108, LM208			UNIT
				MIN	TYP	MAX	MIN	TYP	MAX	
V_{IO}	Input offset voltage	$R_S = 50\,\Omega$	25°C		0.3	0.5		0.7	2	mV
			Full range			1			3	
α_{VIO}	Temperature coefficient of input offset voltage		Full range		1	5*		3	15*	μV/°C
I_{IO}	Input offset current		25°C		0.05	0.2		0.05	0.2	nA
			Full range			0.4			0.4	
α_{IIO}	Temperature coefficient of input offset current		Full range		0.5	2.5*		0.5	2.5*	pA/°C
I_{IB}	Input bias current		25°C		0.5	2		0.5	2	nA
			Full range			3			3	
V_{ICR}	Common-mode input voltage range	$V_{CC\pm} = \pm 15\,V$	Full range	±13.5			±13.5			V
V_{OM}	Maximum peak output voltage swing	$V_{CC\pm} = \pm 15\,V$, $R_L = 10\,k\Omega$	Full range	±13			±13			V
A_{VD}	Large-signal differential voltage amplification	$V_{CC\pm} = \pm 15\,V$, $V_O = \pm 10\,V$, $R_L \geq 10\,k\Omega$	25°C	80	300		50	300		V/mV
			Full range	40			25			
r_i	Input resistance		25°C	30*	70		30*	70		MΩ
CMRR	Common-mode rejection ratio		Full range	96			85			dB
k_{SVR}	Supply-voltage rejection ratio ($\Delta V_{CC\pm} / \Delta V_{IO}$)		Full range	96			80			dB
I_{CC}	Supply current		25°C		0.3	0.6		0.3	0.6	mA
			105°C, 125°C			0.4			0.4	

*On products compliant to MIL-STD-883, Class B, these parameters are not production tested.
†Full range is −40°C to 105°C for the LM208 and LM208A and −55°C to 125°C for the LM108 and LM108A.

LM308, LM308A
OPERATIONAL AMPLIFIERS

electrical characteristics at specified free-air temperature, $V_{CC\pm} = \pm 5\,V$ to $\pm 15\,V$ (unless otherwise noted)

PARAMETER		TEST CONDITIONS	$T_A^{\dagger}$	LM308A			LM308			UNIT
				MIN	TYP	MAX	MIN	TYP	MAX	
V_{IO}	Input offset voltage	$R_S = 50\,\Omega$	25°C		0.3	0.5		2	7.5	mV
			Full range			0.73			10	
α_{VIO}	Temperature coefficient of input offset voltage		Full range		2	5		6	30	µV/°C
I_{IO}	Input offset current		25°C		0.2	1		0.2	1	nA
			Full range			1.5			1.5	
α_{IIO}	Temperature coefficient of input offset current		Full range		2	10		2	10	pA/°C
I_{IB}	Input bias current		25°C		1.5	7		1.5	7	nA
			Full range			10			10	
V_{ICR}	Common-mode input voltage range	$V_{CC\pm} = \pm 15\,V$	Full range	±14			±14			V
V_{OM}	Maximum peak output voltage swing	$V_{CC\pm} = \pm 15\,V,$ $R_L = 10\,k\Omega$	Full range	±13			±13			V
A_{VD}	Large-signal differential voltage amplification	$V_{CC\pm} = \pm 15\,V,$ $V_O = \pm 10\,V, R_L \geq 10\,k\Omega$	25°C	80	300		25	300		V/mV
			Full range	60			15			
r_i	Input resistance		25°C	10	40		10	40		MΩ
CMRR	Common-mode rejection ratio		Full range	96			80			dB
k_{SVR}	Supply-voltage rejection ratio ($\Delta V_{CC\pm} / \Delta V_{IO}$)		Full range	96			80			dB
I_{CC}	Supply current		25°C		0.3	0.8		0.3	0.8	mA

†Full range is 0°C to 70°C.

LM108, LM108A, LM208, LM208A, LM308, LM308A
OPERATIONAL AMPLIFIERS

TYPICAL CHARACTERISTICS†

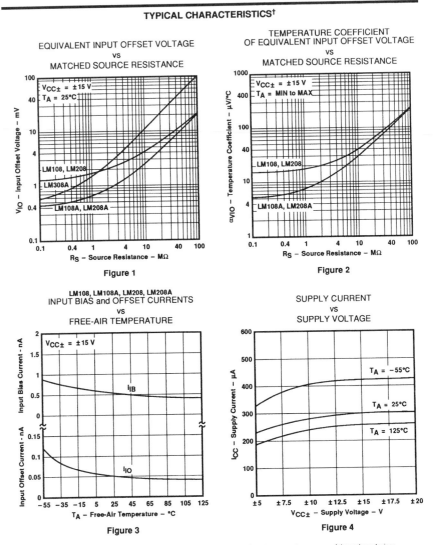

EQUIVALENT INPUT OFFSET VOLTAGE
vs
MATCHED SOURCE RESISTANCE

Figure 1

TEMPERATURE COEFFICIENT
OF EQUIVALENT INPUT OFFSET VOLTAGE
vs
MATCHED SOURCE RESISTANCE

Figure 2

LM108, LM108A, LM208, LM208A
INPUT BIAS and OFFSET CURRENTS
vs
FREE-AIR TEMPERATURE

Figure 3

SUPPLY CURRENT
vs
SUPPLY VOLTAGE

Figure 4

†Data at high and low temperatures are applicable only within the rated operating free-air temperature ranges of the various devices.

LM108, LM108A, LM208, LM208A, LM308, LM308A
OPERATIONAL AMPLIFIERS

TYPICAL CHARACTERISTICS†

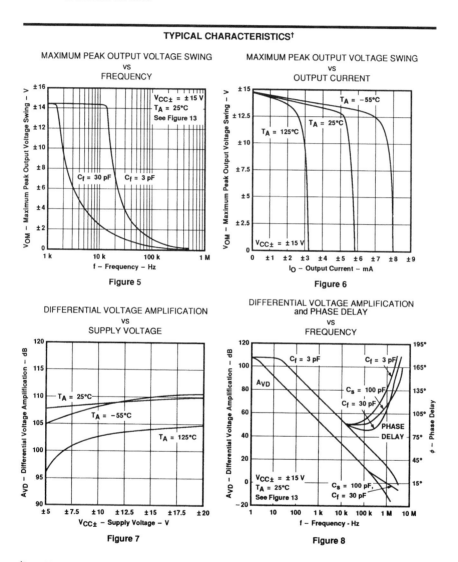

Figure 5

Figure 6

Figure 7

Figure 8

†Data at high and low temperatures are applicable only within the rated operating free-air temperature ranges of the various devices.

LM108, LM108A, LM208, LM208A, LM308, LM308A
OPERATIONAL AMPLIFIERS

TYPICAL CHARACTERISTICS

SUPPLY VOLTAGE REJECTION RATIO
vs
FREQUENCY

CLOSED-LOOP OUTPUT IMPEDANCE
vs
FREQUENCY

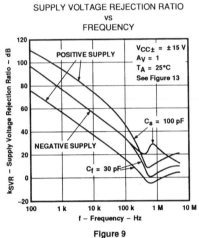

Figure 9

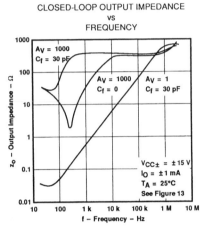

Figure 10

EQUIVALENT INPUT NOISE VOLTAGE
vs
FREQUENCY

VOLTAGE FOLLOWER
PULSE RESPONSE

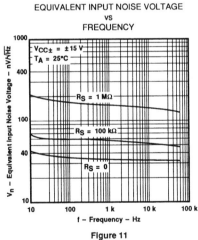

Figure 11

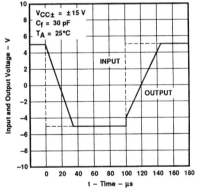

Figure 12

LM108, LM108A, LM208, LM208A, LM308, LM308A
OPERATIONAL AMPLIFIERS

APPLICATION INFORMATION

frequency compensation

Figure 13 shows the frequency compensation circuits for standard compensation, alternate compensation, and feed-forward compensation. The alternate compensation circuit improves supply voltage rejection by a factor of ten.

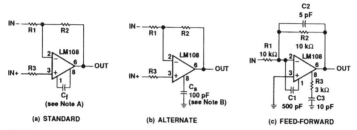

| (a) STANDARD | (b) ALTERNATE | (c) FEED-FORWARD |

NOTES: A. $C_f \geq R1 C_O /(R1 + R2)$, C_O = 30 pF, bandwidth and slew rate are proportional to $1/C_f$.
B. Bandwidth and slew rate are proportional to $1/C_s$.

Figure 13. Frequency Compensation Circuits

input guarding

Input guarding is used to reduce surface leakage (see Figure 14). Both sides of the board must be guarded. Bulk leakage reduction is less than surface leakage reduction and depends on the guard-ring width. The guard ring is connected to a low-impedance point at the same potential as the sensitive input leads. Connections for various op-amp configurations are shown in Figure 15.

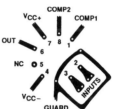

Figure 14. Input Guarding

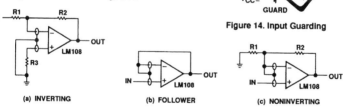

| (a) INVERTING | (b) FOLLOWER | (c) NONINVERTING |

Figure 15. Guard Ring Connections for Various Op Amp Configurations

LM108, LM108A, LM208, LM208A, LM308, LM308A
OPERATIONAL AMPLIFIERS

APPLICATION INFORMATION

Input protection

Current is limited by R2 even when the input is connected to a voltage source outside the common-mode range [see Figure 16(a)]. If one supply reverses, current is controlled by R1. These resistors do not affect normal operation. The input resistor controls the current when the input exceeds the supply voltages, when the power for the op amp is turned off, or when the output is shorted [see Figure 16(b)].

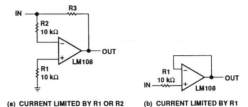

(a) CURRENT LIMITED BY R1 OR R2   (b) CURRENT LIMITED BY R1

Figure 16. Input Protection

Input offset voltage testing

The test circuit for input offset voltage is shown in Figure 17. This circuit is also used as the burn-in configuration with supply voltages equal to ±20 V, R1 = R3 = 10 kΩ, R2 = 200 Ω, AV = 100.

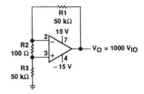

NOTE A: Resistors must have low thermoelectric potential.

Figure 17. Test Circuit for Input Offset Voltage

LM108, LM108A, LM208, LM208A, LM308, LM308A
OPERATIONAL AMPLIFIERS

APPLICATION INFORMATION

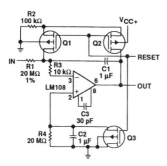

NOTE A: Q1 and Q3 should not have internal
gate-protection diodes.

Figure 18. Low-Drift Integrator With Reset

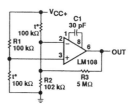

NOTE A: R1 = R2R3/(R2 + R3).

**Figure 19. Amplifier for Bridge
Transducers**

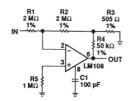

NOTE A: R2 > R1, R2 >> R3,
$A_V = R2(R3 + R4)/R1R3$.

**Figure 20. Inverting Amplifier With High
Input Resistance**

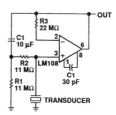

**Figure 21. Amplifier for Piezoelectric
Transducers**

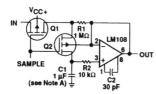

NOTES: A Teflon, polyethylene, or polycarbonate
dielectric capacitor.
B. Worst-case drift is less than 2.5 mV/s.

Figure 22. Sample-and-Hold Amplifier

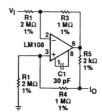

NOTE A: $I_O = (R3)V_I/R1R5$
R3 = R4 + R5
R1 = R2

Figure 23. Bilateral Current Source

LM108, LM108A, LM208, LM208A, LM308, LM308A
OPERATIONAL AMPLIFIERS

APPLICATION INFORMATION

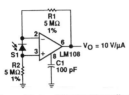

**Figure 24. Amplifier for Photodiode
Sensor**

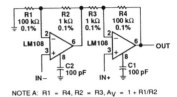

NOTE A: R1 = R4, R2 = R3, A_V = 1 + R1/R2

**Figure 25. Differential-Input Instrumentation
Amplifier**

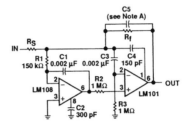

NOTES: A. $C5 = 6 \times 10^{-8}/R_f$
      B. Power bandwidth = 250 kHz
      C. Small-signal bandwidth = 3.5 MHz
      D. Slew Rate = 10 V/μs
      E. The LM101 increases speed, raises high-
         and low-frequency gain, increases output
         drive capability, and eliminates thermal
         feedback.

Figure 26. Fast Summing Amplifier

Appendix 2
Logarithmic amplifier data sheet

Texas Instruments of Manton Lane, Bedford, England, have given their kind permission for the inclusion of the following data sheets in this book:

- Logarithmic Amplifiers – TL441AM

These data sheets are reproduced exactly as they appear in the *Texas Instrument Linear Circuits Handbook*.

TL441AM
LOGARITHMIC AMPLIFIER

D956, JUNE 1976—REVISED FEBRUARY 1989

- Excellent Dynamic Range
- Wide Bandwidth
- Built-In Temperature Compensation
- Log Linearity (30 dB Sections) . . . 1 dB Typ
- Wide Input Voltage Range

description

This monolithic amplifier circuit contains four 30-dB logarithmic stages. Gain in each stage is such that the output of each stage is proportional to the logarithm of the input voltage over the 30-dB input voltage range. Each half of the circuit contains two of these 30-dB stages summed together in one differential output that is proportional to the sum of the logarithms of the input voltages of the two stages. The four stages may be interconnected to obtain a theoretical input voltage range of 120 dB. In practice, this permits the input voltage range to be typically greater than 80 dB with log linearity of ± 0.5 dB (see application data). Bandwidth is from dc to 40 MHz.

This circuit is useful in military weapons systems, broadband radar, and infrared reconnaissance systems. It serves for data compression and analog compensation. This logarithmic amplifier is used in log IF circuitry as well as video and log amplifiers. The TL441AM is characterized for operation over the full military temperature range of −55 °C to 125 °C.

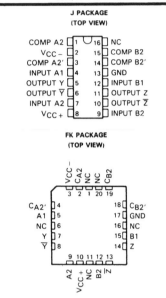

J PACKAGE
(TOP VIEW)

COMP A2	1	16	NC
V$_{CC-}$	2	15	COMP B2
COMP A2'	3	14	COMP B2'
INPUT A1	4	13	GND
OUTPUT Y	5	12	INPUT B1
OUTPUT Ȳ	6	11	OUTPUT Z
INPUT A2	7	10	OUTPUT Z̄
V$_{CC+}$	8	9	INPUT B2

FK PACKAGE
(TOP VIEW)

NC — No internal connection

functional block diagram (one half)

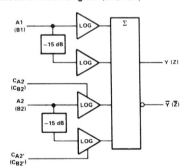

$Y \propto \log A1 + \log A2$; $Z \propto \log B1 + \log B2$
where: A1, A2, B1, and B2 are in dBV, 0 dBV = 1 V.
C_{A2}, $C_{A2'}$, C_{B2}, and $C_{B2'}$ are detector compensation inputs.

TL441AM
LOGARITHMIC AMPLIFIER

schematic

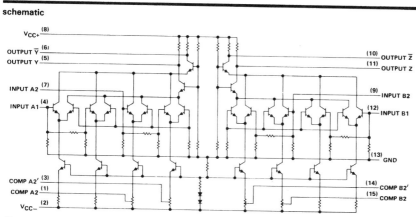

Pin numbers shown are for the J package.

absolute maximum ratings over operating free-air temperature range (unless otherwise noted)

Supply voltages (see Note 1): V_{CC+} .... 8 V
V_{CC-} .... −8 V
Input voltage (see Note 1) .... 6 V
Output sink current (any one output) .... 30 mA
Continuous total dissipation .... See Dissipation Rating Table
Operating free-air temperature range .... −55°C to 125°C
Storage temperature range .... −65°C to 150°C
Case temperature for 60 seconds: FK package .... 260°C
Lead temperature 1,6 mm (1/16 inch) from case for 60 seconds: J package .... 300°C

NOTE: 1. All voltages, except differential output voltages, are with respect to network ground terminal.

DISSIPATION RATING TABLE

PACKAGE	$T_A \leq 25°C$ POWER RATING	DERATING FACTOR	DERATE ABOVE T_A	$T_A = 70°C$ POWER RATING	$T_A = 125°C$ POWER RATING
FK	500 mW	11.0 mW/°C	104°C	500 mW	275 mW
J	500 mW	11.0 mW/°C	104°C	500 mW	275 mW

recommended operating conditions

	MIN	NOM	MAX	UNIT
Peak-to-peak input voltage for each 30-dB stage	0.01		1	V
Operating free-air temperature, T_A	−55		125	°C

TL441AM
LOGARITHMIC AMPLIFIER

electrical characteristics, V_{CC+} = 6 V, V_{CC-} = −6 V, T_A = 25°C

PARAMETER	TEST FIGURE	MIN	TYP	MAX	UNIT
Differential output offset voltage	1		±25	±70	mV
Quiescent output voltage	2	5.45	5.6	5.85	V
DC scale factor (differential output), each 3-dB stage, −35 dBV to −5 dBV	3	7	8	11	mV/dB
AC scale factor (differential output)			8		mV/dB
DC error at −20 dBV (midpoint of −35 dBV to −5 dBV range)	3		1	2.6	dB
Input impedance			500		Ω
Output impedance			200		Ω
Rise time, 10% to 90% points, C_L = 24 pF	4		20	35	ns
Supply current from V_{CC+}	2	14.5	18.5	23	mA
Supply current from V_{CC-}	2	−6	−8.5	−10.5	mA
Power dissipation	2	123	162	201	mW

electrical characteristics over operating free-air temperature range, V_{CC+} = 6 V, V_{CC-} = −6 V (unless otherwise noted)

PARAMETER		TEST FIGURE	MIN	TYP	MAX	UNIT
Differential output offset voltage		1			±100	mV
Quiescent output voltage		2	5.3		5.85	V
DC scale factor (differential output) each 30-dB stage, −35 dBV to −5 dBV		3	7		11	mV/dB
DC error at −20 dBV (midpoint of −35 dBV to −5 dBV range)	T_A = −55°C	3			4	dB
	T_A = 125°C				3	
Supply current from V_{CC+}		2	10		31	mA
Supply current from V_{CC-}		2	−4.5		−15	mA
Power dissipation		2	87		276	mW

PARAMETER MEASUREMENT INFORMATION

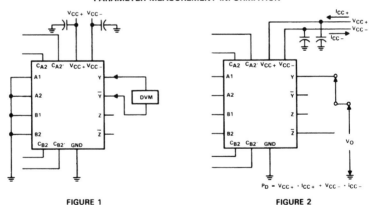

FIGURE 1   FIGURE 2

TL441AM
LOGARITHMIC AMPLIFIER

PARAMETER MEASUREMENT INFORMATION

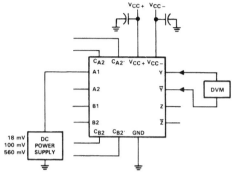

$$\text{Scale Factor} = \frac{[V_{out}(560\ mV)\ -V_{out}(18\ mV)]\ mV}{30\ dB}$$

$$\text{Error} = \frac{|V_{out}(100\ mV)\ -0.5\ V_{out}(560\ mV)\ -0.5\ V_{out}(18\ mV)|}{\text{Scale Factor}}$$

FIGURE 3

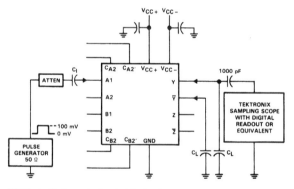

NOTES: A. The input pulse has the following characteristics:
t_w = 200 ns, $t_r \leq 2$ ns, $t_f \leq 2$ ns, PRR $\leq$ 10 MHz.
B. Capacitor C_i consists of three capacitors in parallel: 1 μF, 0.1 μF, and 0.01 μF.
C. C_L includes probe and jig capacitance.

FIGURE 4

TYPICAL CHARACTERISTICS

DIFFERENTIAL OUTPUT OFFSET VOLTAGE
vs
FREE-AIR TEMPERATURE

QUIESCENT OUTPUT VOLTAGE
vs
FREE-AIR TEMPERATURE

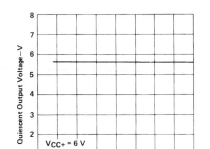

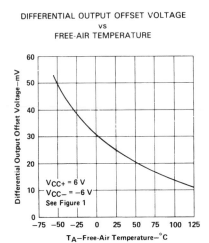

FIGURE 5

FIGURE 6

DC SCALE FACTOR
vs
FREE-AIR TEMPERATURE

DC ERROR
vs
FREE-AIR TEMPERATURE

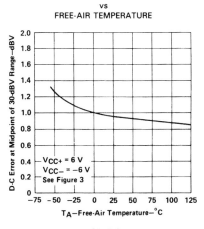

FIGURE 7

FIGURE 8

**TL441AM
LOGARITHMIC AMPLIFIER**

TYPICAL CHARACTERISTICS

OUTPUT RISE TIME
vs
LOAD CAPACITANCE

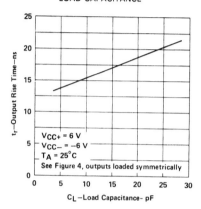

V_{CC+} = 6 V
V_{CC-} = −6 V
T_A = 25°C
See Figure 4, outputs loaded symmetrically

FIGURE 9

POWER DISSIPATION
vs
FREE-AIR TEMPERATURE

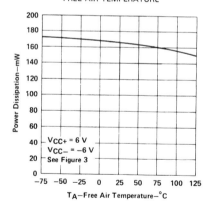

V_{CC+} = 6 V
V_{CC-} = −6 V
See Figure 3

FIGURE 10

TL441AM
LOGARITHMIC AMPLIFIER

TYPICAL APPLICATION DATA

Although designed for high-performance applications such as broadband radar infrared detection and weapons systems, this device has a wide range of applications in data compression and analog computation.

basic logarithmic function

The basic logarithmic response is derived from the exponential current-voltage relationship of collector current and base-emitter voltage. This relationship is given in the equation:

$$m \cdot V_{BE} = \ln \left[(I_C + I_{CES})/I_{CES} \right]$$

where:

I_C = collector current
I_{CES} = collector current at $V_{BE} = 0$
$m = q/kT$ (in V^{-1})
V_{BE} = base-emitter voltage

The differential input amplifier allows dual-polarity inputs, is self-compensating for temperature variations, and is relatively insensitive to common-mode noise.

functional block diagram

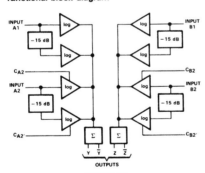

FIGURE 11

logarithmic sections

As can be seen from the schematic, there are eight differential pairs. Each pair is a 15-dB log subsection, and each input feeds two pairs for a range of 30-dB per stage.

Four compensation points are made available to allow slight variations in the gain (slope) of the two individual 15-dB stages of input A2 and B2. By slightly changing the voltage on any of the compensation pins from its quiescent value, the gain of that particular 15-dB stage can be adjusted to match the other 15-dB stage in the pair. The compensation pins may also be used to match the transfer characteristics of input A2 to A1 or B2 to B1.

The log stages in each half of the circuit are summed by directly connecting their collectors together and summing through a common-base output stage. The two sets of output collectors are used to give two log outputs, Y and $\overline{Y}$ (or Z and $\overline{Z}$) which are equal in amplitude but opposite in polarity. This increases the versatility of the device.

By proper choice of external connections, linear amplification, linear attentuation, and many different applications requiring logarithmic signal processing are possible.

input levels

The recommended input voltage range of any one stage is given as 0.01 V to 1 V. Input levels in excess of 1 V may result in a distorted output. When several log sections are summed together, the distorted area of one section overlaps with the next section and the resulting distortion is insignificant. However, there is a limit to the amount of overdrive that may be applied. As the input drive reaches ±3.5 V, saturation occurs, clamping the collector-summing line and severely distorting the output. Therefore, the signal to any input must be limited to approximately ±3 V to ensure a clean output.

output levels

Differential-output-voltage levels are low, generally less than 0.6 V. As demonstrated in Figure 12, the output swing and the slope of the output response can be adjusted by varying the gain by means of the slope control. The coordinate origin may also be adjusted by positioning the offset of the output buffer.

TL441AM
LOGARITHMIC AMPLIFIER

TYPICAL APPLICATION DATA

circuits

Figures 12 through 19 show typical circuits using this logarithmic amplifier. Operational amplifiers not otherwise designated are TLC271. For operation at higher frequencies, the TL592 is recommended instead of the TLC271.

TYPICAL TRANSFER
CHARACTERISTICS

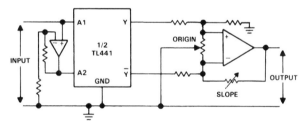

FIGURE 12. OUTPUT SLOPE AND ORIGIN ADJUSTMENT

TYPICAL APPLICATION DATA

TRANSFER CHARACTERISTICS
OF TWO TYPICAL INPUT STAGES

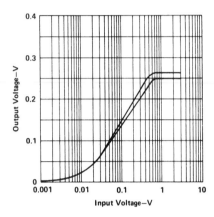

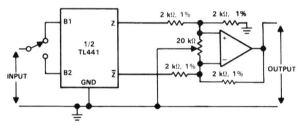

FIGURE 13. UTILIZATION OF SEPARATE STAGES

TL441AM
LOGARITHMIC AMPLIFIER

TYPICAL APPLICATION DATA

TRANSFER CHARACTERISTICS
WITH BOTH SIDES PARALLELED

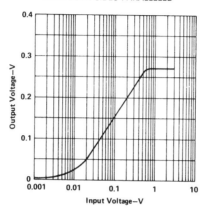

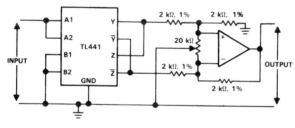

FIGURE 14. UTILIZATION OF PARALLELED INPUTS

**TL441AM
LOGARITHMIC AMPLIFIER**

TYPICAL APPLICATION DATA

TRANSFER CHARACTERISTICS

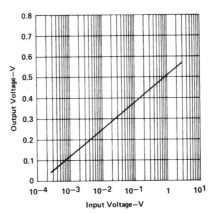

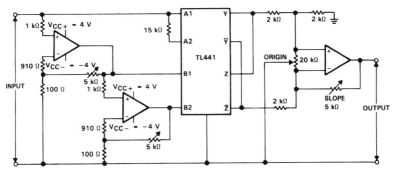

NOTES: A. Inputs are limited by reducing the supply voltages for the input amplifiers to ±4 V.

          B. The gains of the input amplifiers are adjusted to achieve smooth transistions.

Figure 15. LOGARITHMIC AMPLIFIER WITH INPUT VOLTAGE RANGE GREATER THAN 80 dB

TL441AM
LOGARITHMIC AMPLIFIER

TYPICAL APPLICATION DATA

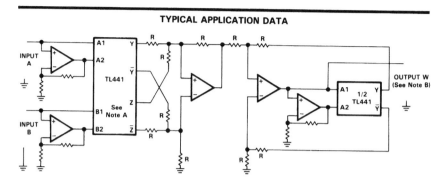

NOTES: A. Connections shown are for multiplication. For division, Z and $\overline{Z}$ connections are reversed.

B. Output W may need to be amplified to give actual product or quotient of A and B.

C. R designates resistors of equal value, typically 2 kΩ to 10 kΩ.

Multiplication: $W = A \cdot B \Rightarrow \log W = \log A + \log B$, or $W = a^{(\log_a A + \log_a B)}$

Division: $W = A/B \Rightarrow \log W = \log A - \log B$, or $W = a^{(\log_a A + \log_a B)}$

FIGURE 16. MULTIPLICATION OR DIVISION

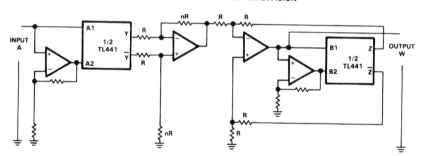

NOTE: R designates resistors of equal value, typically 2 kΩ to 10 kΩ. The power to which the input variable is raised is fixed by setting nR. Output W may need to be amplified to give the correct value.

Exponential: $W = A^n \Rightarrow \log W = n \log A$, or $W = a^{(n \log_a A)}$

FIGURE 17. RAISING A VARIABLE TO A FIXED POWER

**TL441AM
LOGARITHMIC AMPLIFIERS**

TYPICAL APPLICATION DATA

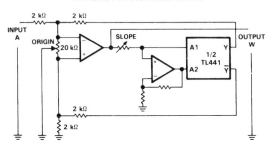

NOTE: Adjust the slope to correspond to the base "a".
Exponential to any base: W = a

FIGURE 18. RAISING A FIXED NUMBER TO A VARIABLE POWER

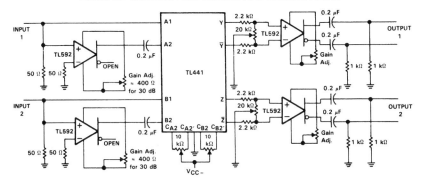

FIGURE 19. DUAL-CHANNEL RF LOGARITHMIC AMPLIFIER WITH 50-dB INPUT RANGE
PER CHANNEL AT 10 MHz

Answers to exercises

Chapter 1

1.1 (a) Yes, $V_x = V_y$; (b) 0.393 V, Y is positive
1.3 R_4 = 120 kΩ, L_4 = 93.6 mH, Z_4 = 120.14 kΩ
1.5 R_4 = 200 kΩ; R_1 and R_2 should be variable 5.3 kΩ to 15.92 kΩ
1.6 0.232 V

Chapter 2

2.1 (a) -10 V; (b) $+12$ V; (c) $+9$ V; (d) -18 V; (e) $+16$ V; (f) $+3$ V; (g) $+28$ mV; (h) -10 V
2.2 (a) -1 V, (b) -60 mV
2.3 (a) R_i = 10 kΩ, R_f = 500 kΩ; (b) 0.024%
2.4 6, 2.5×10^9
2.5 (a) 10 (b) 9.99; (c) 9.0; (d) 4.76; (e) 0.83
2.6 (a) 10; (b) 27
2.7 (a) 15.56 dB; (b) 60 kΩ
2.8 (a) 0.091; (b) 45.4 dB
2.9 1 mA, 0.02%, 50 MΩ
2.10 (a) 200 mV, 84.3$°$ lead; (b) 1.42 V, 45$°$ lead; (c) 2 V, 5.7$°$ lead

Chapter 3

3.2 (a) $-$ 204 mV; (b) $-$ 60 mV
3.4 (a) 5 V; (b) 18.4%

Chapter 4

4.3 178 mV; (a) 166 mV; (b) 190 mV

Chapter 5

5.1 0.1 s
5.2 (a) 1 mV; (b) 31 μV
5.3 (a) 10 μF; (b) -2 V
5.4 A phase inverted, 1 kHz square wave varying between ± 4 V
5.5 250 Hz

Chapter 6

6.1 (a) $+10$ V; (b) $+0.333$ V; -10 V
6.2 1.82 V
6.3 3 V, 2.7 V
6.4 (a) 1366 Hz; (b) $+8.78$%

6.5 The pulse amplitude varies between 0.8 V and 9 V, is of duration 0.355 ms and has a recurrence frequency of 1 kHz

Chapter 7

7.1 3.83 V
7.2 1.5 μF
7.3 Gain = 3.78, $R_2/R_1 = 3.78$, $R_2 = 68$ kΩ, $R_1 = 18$ kΩ
7.4 (a) 11.1 ms; (b) 0.82 V

Chapter 8

8.1 23000 or 43.6 dB
8.2 (a) 9.49 nW; (b) 1.55 μW; (c) 3 mW; (d) 1890 or 32.8 dB
8.3 (a) 31.15 μV; (b) 6.07×10^{-15} W; (c) 7.45×10^{-14}; (d) 2.48×10^{-14} W
8.4 (a) 31.6 mW; (b) 66.67 or 18.2 dB
8.5 4.42×10^{-12} W; 381 or 25.8 dB
8.6 27.78 or 14.4 dB
8.7 455 or 26.6 dB

Chapter 9

9.4 (a) 5.64 MΩ; (b) 23 Hz
9.5 100 kΩ; 200 kΩ
9.6 (a) 395 μV; (b) 310 mV; (c) 477 mV
9.7 10 kΩ, 10 kΩ, 0.32 μF, 7.95 nF
9.8 $A_{obp} = 100$, $f_o = 31.8$ Hz, $Q = 50.5$

Chapter 10

10.1 1.92×10^6 bits/s
10.4 (a) a.m. 4 kHz; f.m. 20 kHz. (b) 36 kHz
10.5 (a) 9 kHz; (b) increases to 36 kHz
10.6 (b) $f = (400 + 50 \sin 100\pi t)$ Hz
10.7 (a) $100 \sin [8\pi \times 10^6 t - 80 \cos 2\pi \times 10^3 t]$ volts;
(b) $4 \times 10^6 + 80 \times 10^3 \sin 2000\pi t)$ Hz; (c) 4.08 MHz, 3.92 MHz

Chapter 11

11.1 (a) 6.4375 V; (b) 0.0625 V
11.2 (a) 10110000; (b) 0.39%
11.3 9.1429 V
11.4 The line carrying I_3 is open circuit
11.5 (a) 14.06 V; (b) 1011_2
11.6 11.11 μs

Chapter 12

12.1 (a) 1110_2, (b) 1100000_2, (c) 10000000001_2
12.2 (a) 0.1_2, (b) $0.00001101...._2$, (c) $11111001...._2$
(d) $11010000....._2$

12.3 (a) 60_{10}, (b) 205_{10}, (c) 273_{10}
12.4 (a) 17H, (b) 77H, (c) 3C4H
12.5 (a) 10110_2, (b) 1010011011111_2, (c) 110000111110010_2
12.6 (a) 39_{10}, (b) 682_{10}, (c) 4369_{10}

Index

8-bit string, 245
a.c. bridges, 5
a.c. amplifiers, 61 *et seq.*
a.m. demodulation, 215
a.m. frequency spectrum, 202
a.m. 198 *et seq.*
a.m. detector, 216
a.m. wave – power distribution, 205
Acquisition time, 136
Active filters, 166 *et seq.*
ADC staircase, 226
ADC and DAC, 221 *et seq.*
Address decoder, 263
Adjacent channel interference, 144
Airspeed measurement, 128
Aliasing, 223
Amplifier slew rate, 38
Amplifier comparator, 99
Amplitude modulation – analysis, 198
Amplitude modulated waveform, 200
Amplitude demodulation, 215
Analogue to digital (ADC) conversions, 221
Analogue filters, 159
Analogue processing applications, 136
Angle modulation, 206
Aperture time, 136
Arithmetic averaging, 130
Astable, 105
Atmospheric noise, 144

Balancing a.c. bridges, 6
Band stop filter, 165
Band-pass filters – state variable technique, 177
Band-pass filter, 162
Bandwidth, 191, 197
Basic linear scaling circuits, 44 *et seq.*
Bessel low-pass filter, 174
Bias current, 38
Binary coded decimal (BCD), 250
Binary number system, 247
Bipolar transmission, 197

Bistable, 105
Bit accuracy, 39
Bode plots, 34, 36, 160
Boltzmann's constant, 147
Boolean algebra, 250
Break frequency, 36
Bridge balance equations, 7
Bridges, 1 *et seq.*
Buffer stage, 21
Buses, 261
Butterworth low-pass filter, 173

Capacitance measurement, 127
Capacitance multiplication, 129
Carrier suppression, 190
Carrier signal, 188
Cauer low-pass filter, 174
Chebyshev low-pass filter, 173
Circuit layout and noise, 146
Clocking pulses, 192
Coded digital pulses, 134
Common mode rejection ratio, 38
Common mode rejection ratio (CMRR), 53
Common mode, 15, 53
Comparators, 97
Complex numbers, 6, 35
Computer controlled process, 264
Counter ramp ADC, 236
Cross-talk, 144
Current-to-voltage converter, 19, 55
Current sources and sinks, 60
Current amplification, 55
Current summation, 57
Current difference to voltage conversion, 57
Current and voltage summing, 20
Cut-off frequency, 161

d.c bridges, 1
Damping, 161, 163
Data transmission link, 189